CLIMATE CHANGE, LAND USE and MONETARY POLICY

THE NEW TRIFECTA

GERALDINE PERRY

Wasteland Press
www.wastelandpress.net
Shelbyville, KY USA

Climate Change, Land Use and Monetary Policy: The New Trifecta
by Geraldine Perry

For more information, or to inquire about special bulk order discounts for fund raising, non-profits or educational purposes, email
geri@thetwofacesofmoney.com

First Printing – December 2013
ISBN: 978-1-60047-927-4
Library of Congress Control Number: 2013955419

Printed in the U.S.A.

0 1 2 3 4 5 6 7 8 9 10 11

Dedicated to the memory of the late Charles Walters, trained economist, rigorous student of history, journalist, author of Unforgiven: The Story of How America Has Exchanged Parity Agriculture for Parity War and numerous other works, monetary reformist and member of the Board of Directors of the American Monetary Institute – and founder of Acres USA. His book Unforgiven, like so much of his life's work, was a compelling argument against the earlier and still extant trifecta of war, land use and monetary policy.

TABLE OF CONTENTS

INTRODUCTION 1

CO2 AND THE FERTILIZATION EFFECT 15

THE WEB OF LIFE AND THE GREAT UNRAVELING........ 45

RETHINKING CARBON SINKS AND SACRED COWS........ 61

CLEAN ENERGY AND PEAK SOIL: QUI BONO? 79

EXPLORING LAND-FRIENDLY ENERGY OPTIONS........ 99

HEEDING NATURE'S ECONOMIC SYSTEM........ 129

THE EARTH BELONGS TO THE LIVING 158

LAND FRIENDLY MONETARY POLICY 190

POSTSCRIPT 221

REFERENCES CITED........ 227

- Chapter 1........ 227
- Chapter 2........ 234
- Chapter 3........ 239
- Chapter 4........ 243
- Chapter 5........ 249
- Chapter 6........ 254
- Chapter 7........ 257
- Chapter 8........ 260

APPENDIX I: WHAT IS MONEY? 262

INDEX........ 269

INTRODUCTION

The debate over global warming continues unabated and unresolved. Extremists, who either serve as unwitting enablers for various business and corporate interests or are seduced by their own economic interests, are often able to dominate whatever discussion is attempted. As a result, fear mongering and hype have been ratcheted up to such a degree that these tactics alone can be used to drown out inconvenient facts, reason and logic.

The entire issue has become so highly charged that any and all information which deviates even slightly from "the party line" is rejected outright. This is unfortunate to the extreme. Not only does it have the effect of shutting down serious discussion but, even worse, it prevents us from correctly analyzing the issue.

The blurring of the term climate change (which refers to long term changes in weather patterns that can be due to natural phenomenon as well as human influence) with the term global warming (which is a type of climate change referring to human-induced increases in surface temperatures brought about by the so-called greenhouse effect) certainly hasn't helped.

The manipulation of these terms has effectively obscured the potentially deadly impact that our carbon-starved soils can have on plant life, and indeed all life. Moreover, our relative ignorance of soil science has paved the way toward a collective fixation on CO2 reduction as a kind of economically viable, substitute plan for dealing head-on with soil carbon depletion, pollution and other forms of land abuse which themselves play key roles in climate change. In short, the blurring of these terms is the very thing that allows us to ignore the manner and extent to which land use policy and practice has affected climate over millennia.

Just one example is Mesopotamia, home of the fabled Garden of Eden and the original epicenter of global trade. Once a boreal forest, old Mesopotamia is now a desert, transformed several thousand years ago by centuries of over-logging, mining, and over cultivating, with much of the associated products destined for foreign markets or devoted to questionable "public projects" such as war and empire building. Land exploitation, born of the need to satisfy the requirements of periodic debt bubbles, did more than merely sustain a burgeoning population. It helped produce the very Cradle of Civilization, the evidence of which was buried long ago under a heavy mantle of sand.

It was in fact the introduction of commodity agriculture and other extractive industries such as logging and mining that led to the legal concept of usufruct, the first examples of which can be found in the Law of Moses and the Hammurabi Code. Usufruct in simplest terms involves the right to the income or use of land without impairment of its substance. Similarly, it was the development of commodity agriculture and other extractive industries that led to the economic treatment of land, and the attendant concept of resource allocation in a finite world.

For millennia, land was understood to be a finite resource, despite the fact that "land" – from which all human needs must come – was also understood to include such entities as the sun and the atmosphere, the soil and the seas, and all the resources that lay within the earth. From this perspective, the economy, and all of mankind, was dependent on the continual exchange and flow of resources within and between each of the many entities that made up the natural world. The economy and the wealth produced within the economy was in other words directly dependent on what finite land can produce. Preserving the right of present and future generations to enjoy the fruits of the land was an important consideration. Destroying the usufruct, and otherwise borrowing against the income of future generations was a thing to be avoided.

Yet, from early Mesopotamia forward, debt has been the dominant feature of monetary systems around the world. To solve the problem of outstanding debt exceeding a given economy's ability to pay, ancient Babylonians came to regard periodic debt jubilees as necessary and normal. While these debt forgiveness cycles helped solve a temporary problem, they did nothing to address the related problem of over-production. Attempts by Mesopotamia and every debt-reliant culture since then to protect the usufruct and "finite land" – while noteworthy – have had to increasingly bend to the extractive characteristics of excess debt.

The Industrial Revolution, with it's illusive promise that technology had made resource extraction, or "land," limitless cast aside lessons that had been learned, and relearned many times, over thousands of years about the crucial need to protect the land. A new, much narrower definition of "land" emerged in economic thinking, eliminating or at least ignoring the natural world's contribution, in economic terms, to man's well being. Concern about impairment of land was no longer a problem because technology had rendered the new, narrower concept of "land" limitless.

Henceforth, only those characteristics which gave land immediate economic value were considered to be of paramount importance in land use policy. Less than three hundred years since the advent of the Industrial Revolution and less than one hundred and fifty years since the introduction of the new economic definition of land, land – in its most inclusive sense – is today visibly exhibiting the disastrous effects of such policies.

It is within the muddled context of climate change, land use, and economic terminologies that so much of what passes for learned discussion about the true nature of our environment, and the extent to which land use and monetary policy impact that environment, falls woefully short. Yet in point of fact, the entire subject of global warming, which has most recently been rechristened as climate change, takes on a critically important, yet largely ignored dimension when viewed through the lens of agriculture – and by extension, land use.

Ignorance of the manner in which land use affects climate and our own health is at the core of why most, if not all, current, so-called global warming solutions simply don't make sense. These solutions are often if not always tech-dependent and are advocated based on their ability to reduce CO2 emissions. They include an array of agricultural-based renewables, "clean" coal and gas, windmill farms, various carbon sequestration projects and a litany of other supposedly earth friendly measures which simply don't stack up when land use considerations are fully taken into account.

In the broadest sense, large scale attempts to reduce CO2 fall under the umbrella of CDR, or Carbon Dioxide Reduction programs, and as such are methods by which to "engineer" the climate. But as the IPCC report of an expert meeting on geo-engineering held in June of 2011 reveals, attempts to sequester CO2 in the ocean, either through changing the ocean's chemistry via fertilization with micronutrients such as iron or by macronutrients such as nitrate and or phosphate – OR by direct injection of CO2, has not been able to solve resultant acidification of the ocean.

Perhaps most egregious of all has been the increased interest in – and deployment of – a smorgasbord of weather modification, geo-engineering, and climate remediation schemes which fall under the SRM or Solar Radiation Management umbrella. These schemes typically involve a variety of toxic chemicals and heavy metals which are injected into the atmosphere via aerosol formats. SRM thus represents a more direct attempt to control the weather and otherwise alter – or "remediate" – climate. One of the problems with SRM programs according to the June 2011 IPCC report was that these types of programs could affect regional precipitation patterns. The major conclusion was that geoengineering itself may constitute "dangerous anthropogenic interference with the climate system."

Suspicion has indeed been growing that weather modification experiments may be disrupting normal weather patterns, interfering with weather forecasting in the short term and leading in the longer term to more persistent disruptions in regional and global weather patterns. Even more problematic, NASA itself has suggested that climate remediation efforts designed to reduce global warming may actually be adding to it.

Of course, it goes without saying that aerosols injected into the atmosphere above our heads eventually but inevitably make their way back to earth. Moreover, the entire process carries with it the extreme likelihood of potentially catastrophic disruptions of normal biological cycles and functions of plant and animal life. Thus, former Vice President Al Gore himself pointed out the obvious on a recent episode of the Ellen Degeneres Show. Paraphrasing Mr. Gore: it's "completely nuts" to geoengineer the climate by pumping still more pollution into the atmosphere, in an effort to modify the warming effects of the sun by blocking it out. Plants, as Mr. Gore correctly suggested, are dependent on the sun, not the chemicals and heavy metals contained within such "sun blocks."

Man has impacted the environment and unfortunately much of this impact has been negative. Widespread, life-destroying pollution is one obvious and appalling result of reckless land use. Additionally, archeology and recent research show that man can and long has dramatically altered his climate through abusive forms of land use practices, practices that primarily involve mining, logging and agricultural activities. A third, less recognized, but monumentally important, result of reckless land use is the progressively greater losses of organic soil carbon, which is one of the key nutrients needed for all plant life, indeed all life.

Soils depleted of carbon produce smaller, less resilient plants which not only give off less oxygen but "breathe in" less atmospheric carbon, more commonly known as CO2. Both soil carbon and CO2 are very much valued by knowledgeable gardeners and farmers for their fertilization effect. Surprising as it may seem to most, it is actually *inadequate* levels of both forms of carbon that are now contributing to declining crop yields. Yet it is our singular obsession with CO2 as THE cause of global warming that prevents us from understanding how it is that too little CO2 might actually increase hunger and starvation.

Plants grown on carbon poor, degraded soils not only provide poorer, less nutrient dense food for animals and humans – but they also require increasing levels of amending, including irrigation and use of various chemicals, the most notable of which are nitrogen fertilizers. Such fertilizers allow us to borrow from soil reserves meant for – indeed required by – future generations, and they destroy soil carbon and the structure of the soil itself. These fertilizers also happen to be heavily reliant on fossil fuels for both production and distribution.

Irrigation, like other modern agricultural practices, causes desertification, or the creation of local deserts, while the run-off from nitrogen fertilizers creates dead zones in waterways via algae blooms that choke off the oxygen supply to aquatic life. Abuse of the land also means that flooding of land and river basins steadily increase as compromised plants and soils become less able to absorb rain and snow melt.

This reduced ability of plants and soil to absorb water is further compounded by the chemical nature of water. Because water has what is known as a polar molecular structure, water molecules with opposite electrical charges attract each other. Water in other words attracts water. This means that when plants and the soil suffer a reduced capacity to absorb water droughts will increase. In sum, weather patterns are altered and the long term result is climate change.

The panoply of feedback loops created by degraded soils is rarely if ever fully examined, nor is the ultimate effect of such soils upon the climate ever seriously addressed within the context of the broader global warming debate. Yet the evidence has long been available that abuse of the land can result in climate change, with Mesopotamia providing one of the best known examples. Today we have a relative handful of academics, including archaeologists, geoscientists, meteorologists, environmentalists, and climate scientists who have begun to devote themselves to this area of study. They are coming to the conclusion that land use and modern agriculture do indeed affect climate and weather patterns.

History provides many lessons concerning land use policy which we can employ to our mutual benefit or destruction according to the kinds of choices we must now make. For example, much of the current transformation of earth's landscape into all manner of food-and-energy plantations is occurring in poor countries located within the Global South on land deemed marginal, i.e. peatland/rain forests that have been inhabited for centuries by indigenous populations who have never held "legal" title to that land. Indigenous peoples are pushed off their lands through partnerships that their governments, in search of revenues, form with big banks, big business and the financially well off.

This is unfortunately a pattern that has plenty of historical precedents, including that of the Native American Indian and small farmers in the American South and West after the Civil War. Wherever and whenever this pattern appears, one has to acknowledge that it not only represents a gross injustice to the people who live in such areas, but in the case of on-going destruction of the world's remaining oxygen-producing rain forests especially, it bodes poorly indeed for the future of the planet.

Importantly, this pattern is intimately tied to a monetary/economic system reliant on debt, as small farmers – especially those in the immediate aftermath of the Revolutionary War and those of the Populist Era – at least partially recognized. As this book reveals, it is this debt-reliant monetary system which, by its nature, demands the construction of increasingly unacceptable legal and economic models for dealing with land use issues.

It is this reliance on debt which drives geopolitical players, led by the big banks and multinational corporations, to exploit the rain forests of the Global South in a manner no less injurious than the commodity style logging and agricultural activities that led to the demise of at least eleven successive empires in ancient Mesopotamia. While war has long been the unfortunate by-product of land and resource exploitation, today land exploitation is an all-too-often accepted by-product of mitigating climate change via questionable CO2 reduction mechanisms.

So it is that great swaths of rain forests are daily raised and waterways polluted in order to develop purportedly "cleaner" burning biofuels as an alternative to oil. To help supply said biofuels, oil palm tree plantations are erected. On the surface they make attractive investments because they serve multiple purposes, including the harvest of trees for lumber and oil seeds, which in turn will be used for food, building and cosmetic products as well as for use as a rich biofuel source. These plantations also provide polluters, as well as ordinary investors, the means by which to either grow or purchase "carbon credits" which of course then allows polluters to keep on polluting.

These kinds of strategies are being employed as commercially viable, if wholly misguided, methods by which to reduce atmospheric CO2 and the world's reliance on fossil fuels. All is occurring as a result of an economic device developed by economists as a method by which industrial pollution might be reduced. Thus, in what may be one of the more unfortunate cases of unintended consequences, CO_2 has itself become the scapegoat for all manner of fossil fuel, agricultural and atmospheric pollution, allowing the globe's largest and most egregious polluters and land destroyers to not only escape unscathed but to reap additional profits through the carbon markets.

So it is that time and again we find that in the increasingly competitive race for revenues and profits, it is not only basic human and environmental concerns but also established facts that get cast aside, rationalized, or misapplied through the singularly short term focus of various business models.

Unfortunately for mankind and the planet as a whole, the natural quest for profits is intensified by what one former central banker by the name of Bernard Lietier described as a deliberate effort on the part of central banks to maintain a chronically insufficient money supply, so as to create incentives for more borrowing. Said borrowing is an essential feature of our current monetary system, since it serves to expand the money supply, the main cost of which is of course thrust onto borrowers.

As the money supply shrinks relative to outstanding debt, debt bubbles form – and eventually burst. These debt bubbles are an accepted, if largely unnecessary, feature of the business cycle. These recurring debt bubbles result in an on-going transfer of real wealth from the hands of producers into the hands of the financial sector, whose main source of wealth is through "unearned" income. Moreover, and as the gap between money available to the real economy and total outstanding debt grows, the drive for profits-at-all-costs intensifies – and the rights and needs of people and the land are increasingly submerged through a complex of legal and political maneuverings.

These factors in themselves point to the urgent need for monetary reform – this reform being that in which the money supply is always kept adequate to the needs of the people and the planet, not the owners of debt. It is in fact our current monetary system that ignores the irrefutable fact that real wealth is wholly dependent upon what the land inclusive can provide. To acknowledge this fact is to understand that a new monetary system is required if the planet is to survive.

It is our collective failure to understand the crucial difference between real wealth and financial wealth that inevitably leads to ill-conceived and planet-destroying land use policies. Just as importantly, it is through our collective lack of understanding of the very real needs of the soil (from which real wealth springs) that we blindly cling, even if only by default, to a wholly misplaced faith in industrial-style agriculture and its endless waves of grain, convinced as we are that this is still the best, and perhaps the only, means by which to feed the world, and now perhaps to meet its energy needs.

This lack of understanding also allows us to ignore the impact of land use on local, regional and global climate. And it allows us to buy too easily into the idea that it is CO2 which needs to be reduced, rather than industrial pollutants from all sources, including agriculture and various "climate remediation" schemes. This lack of understanding also keeps us from seeing how land use policies and practices must be redesigned to reflect the urgent need to protect and nurture nature's life-giving reserves (and thus basic human rights) via a just monetary system. Perhaps worst of all, our lack of understanding of the needs of the land means that we will all too often be cajoled into accepting false solutions which will only serve to keep us on the same trajectory of planetary destruction.

Chapters 1 and 2 lay out in grim detail the environmental and human costs of the current agricultural and land use model, while at the same time deconstructing current assumptions and misconceptions regarding greenhouse gases and environmental pollutants.

Chapter 3 presents a new agricultural model as it is being forged by those who incorporate the needs of the land into their plan. This model, based as it is on nutrient-dense food and not endless amber waves of grain, could dramatically transform the global landscape as millions upon millions of acres of newly established forests are allowed to mature into old growth forests, all while replacing priceless soil carbon and humus *and* producing higher quality, nutrient dense, organic food in amounts sufficient to feed the world 20 or 30 times over. Only this kind of landscape stands a chance of re-establishing the health of soils, humankind and the environment while at the same time putting CO2 back into the soil, thereby mitigating climate change, such as it now exists.

Chapter 4 examines who benefits and who (and what) loses within the context of current "clean energy" projects. Chapter 5 offers energy solutions that are, like the agricultural model discussed in Chapter 3, familiar only to small, but uncommonly dedicated groups of original thinkers whose independent research has led them to swim against the tide of the status quo.

Chapters 6, 7 and 8 explore the implications of the shrinking economic concept of land use together with the legal matrix which surrounds such land use and the relevance of all this to our monetary system. These chapters also investigate the manner in which most of us, including most economists, by and large overlook the pernicious effects that debt dominant monetary systems have on the land and its people. Chapter 8 discusses a template for monetary reform that is not only supported by modern tools and a growing cadre of economists but also centuries of accumulated knowledge. Most importantly it is precisely this kind of template which is absolutely essential to adopt if we are ever to extricate ourselves from the jaws of planetary destruction.

It is hoped that the reader will, as Thomas Jefferson once advised, "fix reason firmly in her seat" and "question with boldness" the received knowledge of the status quo, thereby allowing himself to fully explore and seriously consider what is presented here. It is only through pursuit of these kinds of truly systemic solutions that we will make any headway in saving our precious planet.

Man, owing to an egoism that is too shortsighted about his own interests, his penchant to take pleasure in all that stands at his disposal, in short, owing to his carelessness about the future and his fellow men, seems to labor at annihilating his means of preservation and even at destroying his own species. By everywhere destroying the great plants that protected the soil, for the procurement of objects that satisfy his craving of the moment, he rapidly brings about the sterility of the soil he inhabits, gives rise to the depletion of springs, keeps away the animals who drew their substance from those springs, and causes large parts of the globe, once quite fertile and populated in every respect, to be barren, sterile, uninhabitable, and deserted. Always disregarding the advice of experience, in order to give himself over to his passions, he is perpetually at war with his fellow men, and destroys them from all sides and under all manner of pretext; with the result that one sees populations that were once thriving, become increasingly impoverished. It would seem that man is destined to exterminate himself after having rendered the globe uninhabitable.

Jean-Baptiste de Lamarck (1744-1829) in "Système analytique des connaissances positives de l'homme" (1820)

1

CO2 AND THE FERTILIZATION EFFECT

Few today would deny that Planet Earth is under serious assault and that we ourselves are the perpetrators. The damage to our earthly home has become so plainly visible that it has become impossible to ignore. Our only real choice is either to actively participate in our planet's recovery or be forced to witness our own extinction. Yet the sad truth is that our best efforts to contribute in any meaningful way to earth's rehabilitation are neutralized by controversy, confusion and conflict.

The intensified debate over global warming has only deepened schisms and created still more confusion. But despite all the *Sturm und Drang*, one thing is clear: Our planet is in serious and immediate need of TLC. How much longer she can last is anyone's guess. The big question remains: what to do? Bombarded relentlessly by propaganda from every corner and fed a steady stream of heavily promoted theories built on false assumptions or distorted facts, how are we to know which of the available choices are truly the best ones?

One of our most important choices must involve paradigm shifts in agriculture and its attendant fossil fuel technologies. Strangely, the popular debate over global warming, as it is currently being waged, may become devoid of practical significance once we fully understand the threat that modern agriculture and attendant land use practices and policies pose to planet earth and its peoples. Meanwhile the use of the climate change debate as a political and economic football ensures that millions if not billions of lives will be adversely impacted if current proposed climate change solutions that produce negative impacts on the land and our environment continue to be pursued.

CO2 and other greenhouse gases do indeed loom large, but not in the ways or for the reasons that we might at first assume. Therefore, and for the moment at least, we will say that it matters little whether one counts himself a skeptic or alarmist. Rather, it is industrial-style agriculture and its attendant fossil-fuel related technologies – together with other poorly thought out land use practices and policies – that should be of utmost concern.

Within this context, it matters greatly that we fully understand the manner in which our food system is becoming increasingly compromised on many levels, not the least of which are nutritional, economic, moral, and political. Added to this list are the massively damaging environmental impacts of reckless and ill conceived land use practices which have become so widespread and pervasive that the whole of life on planet earth is seriously threatened. Contrary to popular belief, land use practices are not rooted in human population pressures but rather in a monetary system that has, in the main, been with us since the days of ancient Mesopotamia.

The extreme gravity of the current situation demands that we must be willing to dig deep into the facts, even those that we may have been induced, seduced or bullied into ignoring or denying. Herein begins an in-depth exploration of those facts.

The GHG Debate

Let's begin by briefly stating the obvious. Carbon dioxide, which is the gas held most responsible for global warming, is a natural by-product of the on-going process of respiration in the earth system. In this system, plants and sea plankton in effect breathe in CO2 and breathe out oxygen, while animals, reptiles, fish, and humans do the opposite. In this process plants and organic matter help sequester (or return) carbon to the soil where it serves as a key nutrient required for healthy plant growth. This is what is known as the carbon cycle and it is important to remember that everything has carbon in it, including you and me. CO2 like oxygen is, in other words, a natural and necessary part of the carbon cycle and life itself is dependent on it.

As we all are aware, the climate change debate holds CO2 to be the gas predominantly responsible for potential runaway global warming. But buried in the fine print we find other gases which also are said to play a role. The most familiar of the "forgotten" gases are halo-carbons, methane, nitrogen oxides, and nitrous oxides, along with water vapor and ozone.

While many of these gases can also be man-made, many if not most, including halo-carbons, are produced by natural oxidation processes which take place as organic matter decomposes. (1) As part of these processes, a variety of feedback loops are continually being created as Mother Nature works to maintain equilibrium. It is of course fairly well understood that there is still much to learn about these feedback systems and the manner in which climate might be affected by them.

We are, for example, informed that not only is water vapor the most prevalent gas in the atmosphere but it accounts for most of the greenhouse effect, despite the amount of attention that is currently being focused on CO2. (2) Indeed, according to an article titled "Water Vapor Confirmed as Major Climate Change Player" posted at the NASA website, "Andrew Dessler and colleagues from Texas A&M University in College Station confirmed that *the heat-amplifying effect of water vapor is potent enough to double* the climate warming caused by increased levels of carbon dioxide in the atmosphere." (3) [italics mine] Meanwhile information provided by the National Climate Data Center tells us that "the feedback loop in which water is involved is critically important to projecting future climate change, but as yet is still fairly poorly measured and understood." (4)

Bottom line is that we can say with a fair amount of accuracy that the myriad of effects as well as "the full range of sources of greenhouse gases both natural and anthropogenic is not yet fully understood . . ." (5) One case in point is nitrogen. While not itself typically thought of as a greenhouse gas, it is found in our atmosphere in much greater amounts than CO2. Excluding water vapor, nitrogen comprises 78 percent of Earth's atmosphere, oxygen 21% and carbon dioxide just .038 percent. (6)

We find that not only have earth's atmospheric oxygen levels declined by about one-third (and by 50% in some cities) compared to prehistoric levels (7) but more surprising still "the anthropogenic influence on the N [nitrogen] cycle is *greater than on any of the other biogeochemical cycles.* [As a result of this influence] total nitrogen loading to the global landmass has nearly *doubled* since the pre-industrial era. . . This relatively new input . . . has a *myriad of effects on waters, soils, and the atmosphere.*" (8) [italics mine]

We will return later to the myriad of effects of this anthropogenic nitrogen loading to the global landmass, but for the present let us turn to a 2009 report that revealed a "team of climate scientists from eight U.S. national labs and academic institutions had successfully incorporated the nitrogen cycle into global simulations for climate change for the first time, questioning previous assumptions regarding carbon feedback."

This was important news, said the report, because until now "climate models have ignored the nutrient requirements for new vegetation growth, assuming that all plants on earth had access to as much nutrient flow as needed. By taking the natural demand for nutrients into account, the authors have shown that the stimulation of plant growth over the coming century may be two to three times smaller than previously predicted." (9)

While warmer temperatures, presumably created by additional CO2 remaining in the atmosphere, will actually result in the *increased* availability of nutrients due to accelerated decomposition of plant material, the study found that this process is *not* enough to offset the reduced level of plant growth caused by previously unrecognized nutrient limitations.

Lower levels of plant growth caused by nutrient limitations, the researchers said, means that atmospheric CO2 will then increase over the next century since smaller plants take in less CO2. The economic and political ramifications of this new insight we will examine shortly but more important for our immediate purpose is that these smaller plants will also produce less oxygen for the environment and less, poorer quality food for human consumption.

This study's own conclusion that "potentially significant processes and dynamics are still missing from the simulations" as well as ordinary common sense should tell us that, inasmuch as "the earth's climate system is a highly complex, non-linear system made up of sub-systems that are themselves highly non-linear, [the singular focus on reduction of CO2 emissions] not only oversimplifies the policy questions raised by human GHG emissions, it also misunderstands the significance of the scientific questions." (10) Moreover, and in keeping with the afore-mentioned NASA revelation about water vapor being a major player in global warming, a recent scholarly review of the scientific literature concerning the environmental consequences of increased levels of atmospheric carbon dioxide "leads to the conclusion that increases [of CO2] during the 20th and early 21st centuries have produced no deleterious effects upon Earth's weather and climate." (11)

Nitrogen and Carbon: Farmer's Gold

Because it is their business, farmers and experienced gardeners know that both nitrogen and carbon are needed for plant health – and that soil carbon is needed in significantly larger amounts than nitrogen for soil health, which in turn affects plant health. Essentially "organisms that decompose organic matter use carbon as a source of energy and nitrogen for building cell structure. *[And] They need more carbon than nitrogen.*" (12) [Italics mine] This decomposition process is an important part of what helps build and maintain soil health.

Yet, across the United States and around the world, productive lands are exhibiting troublesome declines in organic matter (humus) and loss of carbon. So critical has the problem of soil degradation become that, by the year 2000, professors Harold van Es and Fred Magdoff, writing for the Sustainable Agricultural Research and Education Program, were declaring this to be one of the major problems the United States, and the world, now faces. (13)

In simplest terms, one could say that Nature created humus to recycle carbon. Thus, humus itself is composed of carbon-containing compounds which originate from living things and then get deposited on or within the earth's elemental structure. Humus not only plays a crucial role in soil fertility and plant nutrition, but loss of humus results in a reduction in the amounts of CO2 that can be respired out of the soil. This is why it can be said that "one of the nutrients that is limiting high yields and causing low profit levels [for farmers] is *inadequate* CO2." (14) [italics mine]

While most of us have been conditioned to believe that reduction of CO2 levels by any means is a good thing, clearly that is not the case with plant and soil health. The truth is that we have somehow overlooked "the dramatic relationships between soil use and climate change. . . It is right here in agriculture where the forgotten chance to save the climate lies." (15)

Rebuilding humus is a time-consuming and location-specific activity with little political, social or financial incentive attached. This may be at least part of the reason why we are so easily persuaded to focus instead on developing tree plantations and similar methods for using plants as "carbon sinks." The carbon sink idea is supported by research which compared the measured amount of CO_2 that actually accumulates in the atmosphere with the measured amount that is released into the atmosphere each year through deforestation and fossil-fuel burning. Essentially, a whole lot of CO_2 goes missing when these two figures are compared.

To illustrate, there are roughly 7.1 gigatons of carbon released each year (GT C/yr) due to the combined effects of deforestation and fossil-fuel burning. But only about half of that is accumulating in the atmosphere, a fact that prompted researchers to investigate where the missing carbon goes. They found that "most of what Nature takes away is absorbed in the surface water of the oceans, which remove about 2.0 Gt C/yr. . . The remaining 1.8 Gt C/yr of anthropogenic CO_2 – what we add minus what is absorbed in the air and oceans – is evidently being removed by increased carbon storage in forests and soils. . ."(16)

It is on this principle that the EPA can state that "in the United States, *since 1990, the management of forests and non-agricultural land has acted as a net sink of CO_2, which means that more CO_2 is removed from the atmosphere, and stored in plants and trees, than is emitted.*" (17) [italics mine] It is also on the carbon sink principle that whole rainforests are being destroyed in order to make way for tree plantations for use as carbon sinks as well as for "green energy" biofuels and other agricultural products

So we see on the one hand that not only does CO_2 serve an oxygen-equivalent function for plants and sea plankton, thereby providing a natural mechanism for sequestering a good portion of the world's "excess" CO_2, but as farmers know, both plants and the soils on which plants depend are positively starved for carbon. However, farmers who employ modern agribusiness methods are involved in a losing battle because much of the carbon-containing humus in agricultural soil is continually lost through over-cultivation and subsequent soil erosion, together with other industrial practices applied to agriculture.

One vivid example of what can happen in situations involving poor land use practices is the Dust Bowl drought that occurred during the 1930's in the Great Plains states, otherwise known as "Breadbasket to the World." Generally regarded by scientists and historians alike to be the worst and most prolonged environmental catastrophe in U.S. history, millions upon millions of acres of the Great Plains were turned into "dust deserts" during the Dust Bowl years.

These dust deserts occurred as a direct result of decades of plowing under and replacing lush and drought-tolerant, moisture retentive prairie grasses with wheat and similarly thirsty annual cash crops. Decades of over cultivation then collided with normal fluctuations in climate. Water vapor that previously had been formed near the earth's surface via respiration of the soil and prairie grasses disappeared and the drier surface air thus lost its natural ability to cool and moderate air temperatures.

Wheat shriveled in the sun, leaving parched and barren dust in its place. Some 480 tons of topsoil *per acre* (along with commensurate amounts of remaining humus and soil carbon reserves) blew eastward as the exposed earth lost all retained moisture. Over a period of more than ten long and agonizing years, dust particles created by wind storms interacted with the sun's radiation and actually made the drought worse and more widespread than it otherwise would have been. (18)

The high plains have never fully recovered from this debacle and some of the land is still sterile and drifting. Today dust storms are returning to the western plains and new questions are being raised as to whether the Dust Bowl may return to wreak new havoc upon the land and the people. (19) Today, dust bowls are not limited to the United States however, for China is now experiencing its own taste of this phenomenon. (20)

To be sure the Dust Bowl taught us important lessons about land management, some of which we have taken to heart. These lessons have not however taught us to scale down or at least approach with caution our overly optimistic reliance on heavy mechanized machinery, or other "miracle" technologies. Instead we have begun to drain the Ogalla Aquifer underlying the Plains States dry through irrigation. And we have tacitly, and through an interlocking network of legal and monetary policy measures, allowed behemoth multinational corporations to thrive and grow by persuading farmers to rely on a wide variety of chemicals including herbicides, pesticides, synthetic hormones, designer seeds – and fertilizers. All of these agricultural land use practices need to be re-examined and many are in truth woefully inadequate short-term solutions to the very serious long term problems of declining soil fertility.

Chemical fertilization is especially detrimental to the soil because humus, and the structure of the soil itself, is destroyed through such means. According to many experts, including agricultural researcher and writer Ken Fousek, not only can we attribute soil carbon decline directly to industrial style agriculture, but we can further say that for every pound of nitrogen fertilizers added over and above what the soil needs there will be the equivalent of 100 pounds of humus destroyed and ten pounds of carbon lost. (21)

Soils destroyed through over-cultivation and chemical band aids such as fertilizers, herbicides and pesticides result in reduced absorption of water from both rain and irrigation. Because less water can be evaporated out of these soils, there is less water moving to the atmosphere which in turn means that fewer clouds will form and less rain will fall. Meanwhile, the water that does not get absorbed by unhealthy soils creates run-off during periods of rain. This run-off goes into gullies, local creeks and streams, and eventually rivers, resulting in not only still more top soil run-off but also flooding problems along the way.

The chemical structure of water molecules also comes into play in the formation of microclimates. Because water molecules have a polar structure with both a partial positive and a partial negative charge, water molecules are attracted to each other like the opposite ends of a magnet. (22) As water evaporates from the flooded rivers and streams of man-made microclimates and from whatever is respirated from parched soils and plants, it collects in the atmosphere. The precipitation that results is often more intense and heavy because the atmosphere surrounding this kind of artificially altered microclimate contains more water.

Within this context, you might want to contemplate the flooding and other extreme weather events that frequently take place in a rough south/northerly direction along the agricultural belts of the Central Plains states or in a west/easterly direction roughly along the coal mining belt of the Southeastern states where mountain top removal is standard practice. Although oversimplified, it is in this general way that weather patterns are altered and climate change is set in motion. In other words, modern industrial-style agricultural practices (along with similar land abuse practices) can and do alter climate, and weather patterns, as science and history clearly shows. (23)

Prevailing farming practices today, like those of the Dust Bowl, actually *reduce* the amount of CO_2 that can be sequestered in the soil, resulting in weakened plants and degraded, depleted soils. Moreover, today's chronic overuse of nitrogen fertilizer, together with tillage and irrigation practices, not only results in significant soil carbon loss, but results in the release into the atmosphere of nitrous oxide which is said to have a global warming potential 310 times that of CO2, while remaining in the atmosphere for one hundred years. (24)

As our earlier discussions of the Dust Bowl and Mesopotamia clearly illustrate, agriculture has long played an important role in regional climate change. Based on photos from NASA, experts say that today two thirds of the world's land surface is threatened by desertification. Yet the land use lessons provided by the Dust Bowl and similar historical examples around the world have taken a back seat to featuring greenhouse gases as star players in the new war against global warming.

But even within the confines of the GHG context, agriculture plays a prominent role since experts estimate that – when all activities related to agriculture, including forestry, are added up – the sector as a whole accounts for nearly 40% of the greenhouse gas emissions which are held to be responsible for global warming. (25) These emissions are predicted to rise by almost 40 per cent by 2030. Importantly, most of the GHG emissions attributable specifically to food production are methane and nitrous oxide, *not* CO2 (which becomes important only when taking into account transportation and refrigeration, packaging and processing, and production of fertilizers and pesticides.) (26)

Methane is produced through a variety of less-than-natural means including oil, gas and coal extraction, landfill sites, waste treatment – including most importantly giant waste "lagoons" attached to comprehensive animal feeding operations or CAFOs, as well as rice cultivation and biomass burning. The main source of nitrous oxide is the production and application of synthetic fertilizer. Not unlike the land use problems of the Dust Bowl days, we see that whatever GHG emissions attributable to agriculture are primarily associated with industrial-style, monocrop agriculture rather than small scale, diversified farming.

In view of our discussion so far, a fundamental overhaul of current regional and national policies and practices as they relate to agriculture would certainly represent a welcome departure from all those legally protected, heavily subsidized plantation-style farms and CAFO operations that blight the landscape and contaminate our air, water and food systems. Ideally, any new agricultural guidelines would focus on reducing *overall* pollution rather than simply focus on a few greenhouse gases. A bonus would be huge reductions in nitrous oxide emissions as well as significant reductions in the amount of methane and various hazardous pollutants that now emanates from CAFO waste "lagoons." Most importantly, such an overhaul would – indeed must – allow and even encourage farming techniques and practices that would serve to improve soil structure and its natural nutrient content overall, including the all-important carbon reserves, thereby sequestering CO2.

Although they may disagree on particulars concerning "organic agriculture," many experts do agree that "the agricultural sector has the potential to mitigate climate change *mainly by increasing the carbon sequestration rate* (i.e. rate at which carbon is stored in the soil) and to a lesser degree, through reduction of some GHG emissions." [italics mine] If workable incentives and polices could be devised in such a way as to get farmers as a whole to actually increase soil carbon, these experts estimate that this approach would be far more successful than that achieved through carbon pricing.

But these same experts also believe that the problems associated with meaningful soil rehabilitation via agriculture are as complex as they are numerous. They include, for example, the need for massive revision of established and currently expanding industrial-style agricultural practices, and the trade policies which protect them, as well as overcoming consumer resistance to higher prices, and "farmer" resistance to lower profits.

Even more importantly, at least within the context of the GHG debate, are the insurmountable problems associated with measuring the amount of carbon stored in the soil. In addition, there seems to be no satisfactory method of rewarding farmers for increasing carbon in their soils; and it is difficult to increase soil carbon in cropped soils, even in no till operations where stubble is retained. The next best thing, the policy-makers agree, is to find ways to price carbon emissions so as to "internalize negative externalities [which in this case is carbon emissions] in the transport, processing, retail and consumer purchase and preparation of food." (27)

What is needed, according to this line of thinking, is to develop tools which will ensure that those who use or produce "carbon-intensive" products that harm the rest of us will "internalize" their choices through additional fees or other penalties, thereby encouraging "better" choices for offending consumers and producers. So it is that the misdirected war on climate change has turned to carbon pricing.

Pyramiding Profits from CO2

Originally, pollutants such as particulate matter, lead, mercury, carbon monoxide and others were at the center of the debate over how best to deal with a clearly ailing planet. Thus cap-and-trade was devised decades ago by economists as a means of reducing what are known as negative externalities, which are the unpaid costs of industrial pollution – unpaid costs being those costs associated with environmental damage which are shifted to the public at large rather than being born solely by the offender.

But heightened concern over the widely held belief in impending and catastrophic global warming has redirected attention away from these types of pollutants and destructive land use practices and towards carbon dioxide and related greenhouse gas emissions. For industries worldwide, this shift in attention has already resulted in obscene and sometimes illicit profits off carbon emissions through cap and trade and the carbon tax.

Simply put, the carbon tax is a tax on the carbon content of fossil fuels that can be placed at various points of production of goods with the hope of thereby reducing CO2 in order to forestall global warming. By way of contrast, the cap and trade system is a tool by which to reduce select air pollutants (to which list CO2 has recently been added), through swaps and credits. In the cap-and-trade system, a cap is placed on a targeted pollutant and permits or credits are either allocated or sold to businesses who are required to hold the number of permits equivalent to their emissions.

Offset credits can also be obtained through "emissions-savings projects" – the vast majority of which are undertaken not only by business but by financial institutions and governments. These projects generate credits which allow pollution over and above the established limit. The offset credits generated by such projects can be swapped or sold through specially designed markets which in theory but not in practice penalizes polluters while rewarding those who can reduce their emissions.

Pronouncements that cap and trade aimed at CO2 reduction is dead in the water may be premature, given the amount of money already invested in such schemes. According to the activist Carbon Trade Watch, the UN's Clean Development Mechanism (CDM) is the largest offsetting scheme with almost 3,000 registered projects in the global South as of April 2011, and over 3,100 further projects awaiting approval. There are, in other words, and as a result of the UN's Clean Development Mechanism alone . . .

> "many land-use and soil carbon initiatives in existence or pending. They rely on dubious 'carbon sequestration' in forests, farms, and soils. They include the Chicago Climate Exchange, the Voluntary Carbon

Standard, the American Carbon Registry, the Agriculture Soil Credit Standard, Oklahoma Carbon Sequestration Certification Program, Climate Action Reserve, the Alberta Offset Market and the US EPA's Climate Leaders.

They all aim to create carbon credits that can be traded internationally as commodities. Most are voluntary initiatives. They will aggregate farmers into large groups and reckon to cover thousands and even millions of hectares, which means the methodologies applied will involve assumptions and large generalizations. . .

Such projects reduce agriculture, food production, people, biodiversity, ecosystems, agricultural biodiversity, land-use patterns and human rights – to carbon for trading in commodity markets. This is presented as the way to help farmers, improve production and tackle climate change. *However, since counting the carbon is too complicated, projects will be based on generalized assumptions over very large areas.* Small producers will be aggregated and managed in big groups, and any project funding is likely to go to the organizers.

It also imposes a top down approach: policies decided by standard setters such as the Voluntary Carbon Standard serving international markets and investors. (28) (italics mine)

To date, a majority of the largest project developers, says Carbon Trade Watch, are financial consultancies, with the majority of credits being purchased for use in the EU Emissions Trading System, or by banks and financial services companies for speculation. (29) However, in January of 2013, the state of California launched its own cap and trade program. Because of California's size, this program is expected to have significant impact on world markets and may yet prove to be a game changer in terms of influencing participation from a heretofore reluctant U.S. government.

The first U.S. federal legislation devised to address pollution was adopted in 1963 and has been revisited and revised periodically since then. The Clean Air Act of 1990 singled out six "common air pollutants," these being lead, sulphur dioxide, nitrogen oxides, carbon monoxide, ozone and particulate matter.

This Act was then amended by the U.S. Congress in order to address those air pollutants that are known, or can reasonably be expected, to cause adverse effects to human or environmental health. One hundred and eighty eight specific pollutants and chemical groups were initially identified as a result of these amendments, including pesticides such as captan, lindane, heptachlor, chlordane, and phosphorous (which is widely used in fertilizers as well as pesticides.) These amendments also modified and extended federal legal authority that had earlier been provided by the Clean Air Acts of 1963 and 1970. However, CO2 and certain other greenhouse gas emissions were not considered "air pollutants" by the EPA until after a 2007 U.S. Supreme Court decision.

The carbon tax, unlike cap and trade, began with the specific goal to reduce carbon dioxide emissions on the premise that CO2 is a "common air pollutant." Thus, as previously stated, the carbon tax is a tax on the carbon content of fossil fuels and can be placed at various points in production. The amount of the tax, which is essentially a price placed on carbon-intensive businesses and their products, can be changed should carbon dioxide exceed acceptable levels. The U.S. federal government has yet to enact a carbon tax, but a handful of states have been experimenting with it. Pricing carbon emissions in this way is not only considered "sound economics" but a Congressional Research Service report maintains that such a tax could eventually provide enough revenue to cut the country's budget deficit (though not its debt) in half.

The U.S. federal government was given the green light to regulate greenhouse gas emissions by way of the 2007 U.S. Supreme decision mentioned earlier. In *Massachusetts vs EPA*, the Court ruled that carbon dioxide and other greenhouse gas emissions are air pollutants under the Clean Air Act. This gave the EPA the power to regulate CO2 and other GHGs as air pollutants if they so chose. The Court additionally ruled that the EPA must first review the scientific evidence to see if such regulations are needed to reduce global warming.

In 2009 the EPA concluded (somewhat curiously given what has been discussed earlier in this chapter) that CO2 and other GHGs threaten the public health and welfare. Regulations followed, beginning with setting minimum standards for renewable fuels, the establishment of a program for greenhouse gas reporting, and the incorporation of GHGs into its pollution permit program. The EPA's finding and its new regulatory program was upheld in the U.S. Court of Appeals for the District of Columbia in 2012.

Also in 2009, the American Clean Energy and Security Act was passed by the U.S. House of Representatives, but not the Senate. That Act would have established a carbon emissions trading program similar to what had already been introduced in the E.U. To date, only state and local governments have taken it upon themselves to experiment with either a carbon tax or cap and trade program. But both programs have garnered a fair amount of attention, in no small part due to the fact that they provide much needed revenues.

The first nationwide cap and trade program was launched as part of the Acid Rain Program in Title IV of the Clean Air Act of 1990. In addition, a regional NoX Budget Trading Program was employed in the Northeast. Two air pollutants were targeted: sulphur dioxide and nitrogen oxide emissions from fossil-fuel fired power plants. The professed success of these programs ignited global interest, soon to be enhanced in no small part by the anticipation of an emerging global carbon emissions market and with it the formation of the carbon industry wherein carbon emission permits are packaged as tradable credits and various forms of derivatives.

It was upon this anticipation that the EU Emissions Trading Scheme was launched in 2005. But almost immediately problems began to surface. By late 2012 the European Commission estimated that 1.7 billion tons of excess offset credits were on the market. This oversupply, as well as the economic downturn were two factors said to have caused the price of offset credits, or "pollution rights" to plummet. While price swings usually provide a welcome opportunity for profit taking, the extreme price drop has made the market far less attractive to finance professionals. Moreover, cheap credits has allowed, for example, brown coal to win out in the competition with other energy sources. But this is not all that has come out of the EU trading scheme.

For example, a veritable flood of highly speculative projects designed to allow European companies to simply calculate away about 20 percent of their emissions mushroomed almost overnight. As a result, a whole new industry quickly grew to a volume of $90 billion, "an industry complete with certifiers, forecasters, dealers and hackers, who trafficked in certificates and created more and more absurd project." This new industry, together with the carbon intensive energy sector has profited immensely from the EU's carbon trading scheme. For these and other reasons, charges have been leveled that "emissions trading seems tailor-made for the energy sector" and that the entire scheme is nothing more than "industry interests designed as climate protection." Need we wonder why it is that there are those who do not want to change it? (30)

Essentially, cap and trade and the carbon tax are market-based solutions to land use problems which themselves have their origin in monetary and legal policies designed to support the large corporation at the expense of small entrepreneurs, the rights of sovereign citizens, and most crucially, the environment upon which we all depend. (31) However inadvertently, both the carbon tax and cap-and-trade make CO2 into the scapegoat for a wide range of pollutants affecting land, air and water resources, while virtually ignoring all manner of land abuse practices and the coming reality of Peak Soil, which is the point at which our soils will no longer be able to produce. Need it be repeated that all of these carbon pricing mechanisms are based on incomplete modeling systems which hold CO2 responsible for global warming. (32)

Barring much-needed monetary reform, which will be discussed in Chapters 6 through 8, and based on what has been discussed so far, one might ask: Rather than focus on carbon emissions, wouldn't it make more sense to place a tax or some other penalty on all forms of true industrial pollutants, starting at the raw materials stage, while at the same time removing subsidies and legal protections for polluting industries? Such an approach might also specifically target certain agricultural land practices, including dependence on soil carbon destroying monocrop agriculture and soil carbon destroying, nitrous oxide emitting nitrogen fertilizer together with other chemical band-aids. Administered fairly, such an approach might also take into consideration the type and amount of land use and pollution that is connected to all currently promoted energy choices, including so-called green energy alternatives.

Instead, our single-minded obsession with CO2 as a common air pollutant means that – according to the Climate Commission – some 33 countries and 18 subnational jurisdictions will have some sort of carbon pricing scheme in place by 2013, imperfect a tool though it has already proven to be. (33) In essence, both cap and trade and the carbon tax are ways to price carbon, and hybridized instruments incorporate elements of each. But placing a price on atmospheric carbon (CO2) essentially creates a system for privatizing the atmosphere by turning CO2 into a commodity – the financial instruments of which can then be traded in the carbon markets.

One of the many problems associated with the carbon tax is that polluting firms would be automatically incentivized to simply move operations to carbon tax free countries, where carbon and many other forms of pollution would continue unabated. It is also a tax whose burden falls disproportionately upon the poor. Similarly, the offset credits obtainable through cap-and-trade provide a means by which polluters can in effect buy their own permission tickets to continue business as usual, and pass the cost of offset credits on to consumers and others downstream from the supply chain. Under this system, investor coalitions also find many reasons, mostly economic in nature, to develop all manner of questionable projects, including tree plantations and similar large monocrop operations by which to grow carbon sinks – and then sell carbon credits obtained from those sinks.

It goes without saying that it is the largest players who have the resources needed to game the system in these and other ways. For example, many of these large players are among those mining the gold that is thought to lie within the carbon derivatives markets. Unbeknownst to the public at large, many large players are also entering into the unchartered waters of climate remediation.

So far the climate change debate has focused on the most cost-effective means by which to achieve CO2 reduction. But particularly in view of the fact that much of the developing world is on the edge of highly speculative, investor-fueled hypergrowth, CO2 reduction isn't coming quickly enough to satisfy many high profile U.S. academic, military, scientific and corporate interests, who smell potentially huge profits in geoengineering for their institutions and companies.

Undoubtedly seen by some as the perfect marriage between corporate capitalism and climate remediation, these experiments with our atmosphere and Earth's climate systems are occurring under our noses and without our consent or opportunity for debate. An online map, constructed July 2012, shows the expanding scope of research and experimentation already underway, and gives but a minor inkling of the kind of massive pollution already taking place in the name of climate remediation. (34)

As mentioned in the Introduction, most current climate remediation schemes fall under the SRM or Solar Radiation Management umbrella. As such, they attempt to reduce the amount of sunlight reaching the Earth by way of injecting salts, sulphur, aluminum, gases and other aerosols into the atmosphere. These aerosols form what is known as "white haze" or man-made clouds which in turn result in clouds with larger water droplets due to the seeding effect of aerosol particulate matter.

As Rosalind Peterson of the Agricultural Defense Coalition points out: "Geoengineering proponents do not want to discuss the man-made clouds produced by aviation and rocket sources [via geoengineering experiments]. They clearly do not want you to know that water vapor is a greenhouse gas that is produced by aviation sources or that it is having a negative impact on our environment." Even worse says Peterson, NASA research studies show that these man-made clouds are actually "trapping warmth in the atmosphere and exacerbating global warming." (35) In view of this revelation it is not entirely without merit that some skeptics posit that those with vested interests in the continuation of these schemes are knowingly helping create the very global warming they purport to "remediate."

Meanwhile, as the atmosphere gets privatized, polluted and engineered via the carbon trade, not to mention climate remediation, so too do formerly communal lands of indigenous populations, resulting in massive land grabs and other forms of human and environmental abuse. All rooted in fears of imminent global warming catastrophe and the presumed need to measure and thereby reduce excess CO2 emissions.

The cap and trade business has been the premier platform for these land grabs, providing both the agribusiness and energy sectors with brand new profit streams in the form of man-made carbon sinks whose value is derived from the well-recognized fact that trees and other plants can (if healthy) sequester, or sink, a significant portion of the CO2 that misused land and various carbon-based technologies emit.

As always, unintended consequences have begun to pile up and yet are pushed aside by the force and motion of the financial markets. For example, and fortuitously enough for investors, financiers and polluters alike, these man-made carbon sinks have been given the legal green light by which to "grow" offset credits which can then be traded or sold through the carbon markets. Well-heeled polluters wishing to escape financial or regulatory repercussions can simply invest in one or more of these carbon sinks in order to obtain offset credits, or simply purchase credits from those who may have them.

An added bonus within the current monetary and legal construct is that these carbon sinks also provide raw materials for the highly problematic but profitable biofuels market, as well as logging and food products. Similar to many CO2-reducing projects, one of several problems concerning biofuels that no one mentions is that "when biofuels are burnt, the carbon dioxide that the plants absorb when they develop in the fields, is returned to the atmosphere. Thus the costs of the CO2 emissions are ignored as also is the [carbon] emission from fertilizers and pesticides used in the harvests, the use of agricultural machinery, the processing and refining, the transport and the infrastructure for distribution. For the researcher, the extra energy costs and carbon emissions are even greater when the biofuels are produced in one country and exported to another." (36) All this at the very same time that these soils are literally being mined – and destroyed – for short term gain.

In this and many other ways, industrial polluters get a free pass while privatizers of the earth's atmosphere and communal lands allow themselves to believe that the carbon business is the way to save us from global warming. Perhaps even worse, these new carbon sink profit centers are built on a clearly failing agricultural model, which is the very same model that forms the basis for predictions that GHG emissions attributable to agriculture will rise by almost 40 per cent by 2030.

Given the money involved and our collective, colossal failure to fully understand the looming crisis in land use, it is no surprise that oil-seed tree plantations have become a favorite in the carbon trade business among corporations, banks and governments alike. Invariably it is the big players, supported by an enabling public, that are both the economic winners of and the driving force behind these ventures. Small farmers, the poor, and the environment are rarely if ever a true part of the equation. (37)

That is why, despite the environmental and human havoc that is closely associated with these man-made carbon sinks, the carbon business (at least until recently) was expected to reach $3 trillion by 2020. Apparently no one was listening when Patrick Birley, the Chief Executive of the European Climate Exchange, announced that carbon trading "doesn't reduce a single tonne of carbon going into the atmosphere." (38)

It certainly helps that these tree plantations have multiple profit streams, which include their use as carbon sinks and their production of logging and food products – not to mention biofuels as a green energy option. Blinded by the lure of potential profits and ensnared by mountains of speculative, highly leveraged investments, big geopolitical players are heavily promoting plantation style biofuels as an economic solution for poor nations and a green alternative for rich nations, declaring that "the world is entering a new era of participation by emerging market nations in global green energy production for transport fuels."(39) And so, while Planet Earth awaits the day when carbon dioxide emissions will drop to officially prescribed levels, the climate remediation specialists promise to keep us all in business.

We have been very amply informed by official sources that "while gases such as carbon dioxide occur naturally in the atmosphere, [it is] through our interference with the carbon cycle [that] we artificially move carbon from solid storage to its gaseous state, thereby increasing atmospheric concentrations." (40) Carbon sinks notwithstanding, we might ask: what are the ways by which we artificially move carbon from solid storage where it resides within and beneath earth's oceanic and vegetative cover?

We can start with mountain top removal for mining coal plus underground exploration and extraction from the earth of oil, gas, coal, copper, gold, platinum and so forth – not to mention the ceaseless creation of a multitude of petrochemicals and industrial land fills, which all contribute to the destruction of our waterways and atmosphere. Last but by no means least we have carbon destroying soil cultivation techniques, along with the clear-cutting or burning of tropical forests and peatlands for installation of tree plantations and other plantation-style monocrop agriculture.

The same loss of soil carbon reserves that takes place in ill-considered forms of soil cultivation (and all mining activities) also occurs through clear-cutting and replacement of old-growth forests with tree plantations, an activity that has been hastened by the carbon trade. Human misery and life-destroying pollution are unaccounted for by-products in the creation of these carbon sinks.

In South America and elsewhere around the globe, for example, one must add the often illegal population-displacement problem associated with the construction of tree plantations to the attendant, worldwide use of the so-called "innocuous" herbicide glyphosate (Round Up) as a means of eliminating competing vegetation. Hardly innocuous, glyphosate consistently manages to kill local wildlife and the fish downstream of application, and "is causing negative impacts on soil and plants. . ." (41) (42) (43) (44)

One has to ask, does it not stretch the bounds of logic to suggest that replacing complex old-growth forest systems with tree plantations (and other plantation-style monocrop agriculture) devoid of "competing vegetation" will solve the "excess" carbon emissions problem on the presumptuous claim that said plantations can serve as carbon sinks? Luckily and should we not trust our own common sense, we now have experts whose research "concludes that [old growth] forests are far better than plantations at ridding the air of carbon dioxide. . ." (45)

The Fertilization Effect

This leads us to a fascinating corollary to the carbon story involving the scientific phenomenon known as the fertilization effect which takes place as a result of increased levels of atmospheric CO2 and which is "greater at warmer temperatures than at cooler temperatures." (46) This fertilization effect is said to be helping the terrestrial biosphere (vegetative areas) to increase CO2 uptake. Why? "[Because most, though not all] plants grow faster in air that contains more CO2. They also retain water more efficiently, since they do not need to open the pores, orstomata, on their leaves as wide to ingest the CO2 that is needed for photosynthesis." (47)

One activist group has produced a video, viewable on line, of an experiment demonstrating the fertilization effect of increased levels of carbon dioxide on plant growth. (48) The video is a visual confirmation of the statement in the above paragraph and is itself factually accurate. The experiment compared one plant-growing chamber that was supplied with 450 ppm CO2 to another plant chamber that had CO2 levels at 1270 ppm.

For points of comparison in this experiment we are informed that the current level of atmospheric CO2 is roughly 400 ppm – and that 450 ppm is the threshold at which the world will theoretically enter into a potentially cataclysmic global warming cycle. Interestingly, 1270 ppm is the level most greenhouse nurseries would prefer – and the level at which much of life on earth evolved. (49)

The results of the experiment speak for themselves. The plant raised in the carbon-rich chamber weighed 44% more, attained greater height, had more leaves and far more roots. Significantly, "this plant not only grew faster and better, but its better root system would make it more drought and heat tolerant." It follows that this larger, healthier plant will also be able to "breathe in" more CO2.

It is of course a bit simplistic to think that earth's atmosphere behaves in precisely the same way as that provided in a greenhouse. This is because the structure of a greenhouse serves as a real physical barrier to air convection while earth's atmosphere actually facilitates convection, however much the so-called glass ceiling created by greenhouse gases may be imagined to serve the same function as that of a greenhouse. Suffice it to say – and as gardeners and farmers well know – the fertilization effect of CO2 is valid, and crops are absolutely starved for both atmospheric and soil carbon because of industrial-style agriculture and other form of land abuse.

Despite the severe nutrient limitations created by carbon starved agricultural soils, a group of U.S. scientists recently came to the astonishing conclusion "that the increase in agricultural yields between 1961 and 2005 cut greenhouse gas emissions overall, despite higher emissions from factors such as fertilizer production and use."

These scientists concluded that "intensive agriculture has a bad rap for being environmentally unfriendly . . . We find that although intensive agriculture has resulted in increased emissions from fertilizer production and application, the amount of land spared from conversion to cropland by higher yields dominates." (50)

This emphasis on greenhouse gas reduction via the purportedly higher yields supplied by industrial agriculture completely ignores the numerous and widespread negative effects of said agriculture on the land and its seriously depleted carbon stores. It also ignores the manner in which air and water quality is damaged or even destroyed as a direct result of such agriculture. And perhaps most ominously it ignores the declining nutritive quality of foods that are grown on such soils. Yet, however ill-conceived this prevailing view may be, it does provide ammunition for those with vested interests in maintaining the status quo, particularly within the framework of the current GHG debate.

This status quo, unfortunately for human kind and the future of the planet, would have us all believe that industrial agriculture and its new carbon sink/biofuels profit streams are perfectly justifiable within the context of global warming and attendant GHG measurements, never mind the earth-damaging effects of said agriculture and attendant land use policies.

Food Quality Depends on Soil Quality

Even setting aside the long and growing list of very serious environmental concerns of modern industrial-style agriculture, the higher crop yields achieved through it are proving to be sorely lacking in their ability to nourish the world, both in terms of food security and food quality, however spectacular overall yields may be.

For example and echoing concerns raised in Europe, England and the American colonies many generations earlier and experienced by ancient cultures whose civilizations collapsed when their soils gave out due to intensive forms of agriculture and clear-cutting, data collected by the USDA and England from the early 1900's forward shows a progressive decline in both animal health and soil nutrients, mostly due to soil carbon depletion resulting from over-cultivation. (51) (52) (53) (54) (55) (56)

In 1936 the USDA submitted a report to Congress detailing the severity of the health-and-nutrient-loss problem. U.S. Senate Document 264 of 1936 reflected the USDA's report with the following statement: "The alarming fact is that food . . . now being raised on millions of acres of land that no longer contains enough of certain needed nutrients, are starving us no matter how much we eat of them."

In response, Congress in 1939 mandated that trace amounts of six synthetic, unstable nutrients (out of the original 100 organic ones) be added back to heavily processed, denatured flour made from USDA-verified nutrient-depleted wheat. At the same time, use of chemical fertilizers, which had been introduced in England, Europe and the United States during the mid-1800's, nearly doubled during the 1940s and yields skyrocketed. The "green revolution," with its heavy use of fossil-fuel inputs, mechanization of planting and harvesting activities, as well as the development of genetically modified, high yield, designer crops specifically suited to industrialization, had begun.

In the early 1960's Congress was informed that every mineral that had previously been present in our food, except the three used to fertilize modern farmlands, was anywhere from 10 to 40 percent lower than it had been 60 years before. More recently, a global review of science-based research into the impact of environmental factors on intelligence found that the intelligence of millions of people around the world is being harmed not only by environmental pollutants such as lead, PCBs and radiation from uranium mines and nuclear power plants, but also by the loss of micronutrients like iron and iodine through soil erosion which then impoverishes food crops.

What is even more striking about this review and as pointed out by its author, Dr. Chris Williams – who is a social scientist at the Institute of Education, London University, "one problem could compound another, with iron deficiency in children, for example, able to increase their lead uptake. We only have single-substance science, which does not account for compounding effects. So the overall scale of the problem is far greater than previously estimated." (57)

Today, as the mis-named, mis-understood green revolution continues to expand its legal and economic stranglehold across the globe, mankind is being forced to learn or relearn, inch by painful inch, that crop yields have little to do with crop quality, food security for the global poor – or human nourishment. Moreover and given the fertilization effect that CO2 exerts on plants, the single-minded reduction of atmospheric CO2 only stands to make this problem of crop quality and food security even worse – even as it opens the door for ill-advised climate remediation and weather modification experiments which add to the pollution load of the environment, and if NASA is right, even contribute to climate change.

Nature herself has of late been forcing more and more of us to see with our own eyes and learn through our own experiences that crop yields within the agribusiness model have everything to do with the systematic degradation of earth's systems – including earth's critically important soil carbon stores. This degradation is the result of a mixture of highly questionable land management practices, including both age-old and modern cultivation and clear-cutting practices, which themselves are now being recognized by various researchers as a primary factor in both local and global climate change. (58) (59)

Included within this degradation of the land are the sulphur dioxides and nitrogen oxides emanating from fossil fuels which together create life-destroying acid rain. No less important is the massive overuse of nitrogen fertilizer, which is the main source of anthropogenic nitrous oxide, a greenhouse gas with a global warming potential 296-310 times that of CO2, while remaining in the atmosphere for one hundred years. Then we have the lab-created panoply of deadly VOC-based fertilizers and pesticides which must be partnered with patent-protected, enormously profitable GMO crops designed to withstand anthropogenic chemicals. These are the kind of negative externalities which we need to address.

Global warming notwithstanding, one does have to wonder what will happen once these highly speculative, hugely profitable innovations no longer work their illusory magic on thoroughly lifeless soils. We can only use so many parking lots. Within this context we may soon be forced to recognize, as academics are beginning to do, that worries over CO2 reduction may have to take a back seat to land use policies, which themselves may contribute far more than most of us assume to long term climate change, not to mention human health. (60) (61)

2

THE WEB OF LIFE AND THE GREAT UNRAVELING

In 1936, the USDA submitted a report to Congress concerning the dire state of American farm soils. A few months later U.S. Senate Document 264 summarized that report with the following statement: "The alarming fact is that food . . . now being raised on millions of acres of land that no longer contains enough of certain needed nutrients, are starving us no matter how much we eat of them."

That was in 1936. Very shortly thereafter, the impressive gains in crop yield made by the subsequent green revolution seduced many to turn their attention away from the real problem, despite constant warnings from eminent soil scientists such as William Albrecht and many others. So it was that by the early 1960's Congress was given more bad news about the condition of our soils when they were informed that every mineral, except the three used to fertilize modern farmlands, was anywhere from 10 percent to 40 percent lower than it had been 60 years before.

Since that time and as the misnamed green revolution quickly spreads across the globe, soil quality everywhere has only gone from bad to truly appalling, a fact that has been corroborated decade after decade through volumes upon volumes of research.

Without question our tendency to accept temporary quick fixes while ignoring systemic issues has put us on the fast track to our own demise. And just as the highly profitable green revolution pushed aside well-considered objections to its inherent problems, so too has climate change and its ill-conceived, highly profitable solutions obscured the potentially catastrophic damage that industrial agriculture is doing to planet earth and its peoples. As a result, an unholy alliance between Big Ag, Big Energy and climate change has been forged, and it is to our detriment that we ignore it.

Shifting Blame

An abbreviated discussion of environmental degradation begins with the most prominent chemical groups of soil contaminants. These include fuel hydrocarbons, PCBs, and pesticides, all used extensively in modern agriculture-related activities. PCBs, by the way, are used as pesticide extenders and are part of a class of compounds known as organochlorines – of which the super-toxic dioxin is also a member.

High concentrations of dioxin occur primarily in agricultural settings where herbicides and pesticides are used, as well as near the industrial sites where they are produced. Alarmingly, dioxins that are formed during the burning of household plastic products (as well as the burning of industrial and medical waste) can be carried for hundreds of miles on specks of ash where they settle on crops and pastures, or migrate into streams, lakes and the ocean. Eaten by fish and livestock, they then make their way up the food chain and accumulate in the fat cells of our bodies.

Inorganic or non-carbon containing compounds used in agriculture include nitrates and phosphates used in fertilizer applications among other contaminants. Both carbon-containing and non-carbon containing chemicals can leach into groundwater, and also contaminate streams, lakes and oceans through run-off.

It goes without saying that health can be adversely affected by direct contact with contaminated soil or by way of inhalation of chemicals which have vaporized out of the soil. Nitrogen, a form of which is used ubiquitously in fertilizers, "escapes through ammonia volatilization and denitrification. . . Sewage sludge [often used in fertilizer applications, and as distinct from healthy manures] contains heavy metals and, if applied repeatedly or in large amounts, the treated soil may accumulate heavy metals and consequently become unable to even support plant life. In addition, chemicals that are not water soluble contaminate plants that grow on polluted soils, and they also tend to accumulate increasingly toward the top of the food chain." (1)

Five of the six common air pollutants listed by the EPA pursuant to the Clean Air Act of 1990 (2) – lead, sulphur dioxide, nitrogen oxides, carbon monoxide, and particulate matter together with numerous volatile organic compounds or VOCs – are all produced through industrial processes involving fuel combustion. These would include power plants, transportation and a wide variety of agribusiness activities. To this is added the aerosol fallout resulting from widespread climate remediation experiments now underway. All of these activities produce massive amounts of common air pollutants which seriously impact air, water and soil quality.

Scientists, for example, have confirmed that sulphur dioxide and nitrogen oxides are the primary causes of acid rain. Even worse, "acid deposition penetrates deeply into the fabric of an ecosystem, changing the chemistry of the soil as well as the chemistry of the streams and narrowing, sometimes to nothing, the space where certain plants and animals can survive. Because there are so many changes, it takes many years for ecosystems to recover from acid deposition, even after emissions are reduced and the rain becomes normal again." (3)

Thus, while a 2012 report to Congress suggested that the Acid Rain Program established through an amendment to the Clean Air Act had indeed resulted in substantial reduction of both sulphur dioxide and nitrogen oxides, it also held that more work needs to be done. For example, and "despite these emission reductions, the report indicates that full recovery from the effects of acid rain is not likely for many sensitive forests and aquatic ecosystems. For example, in the Adirondack Mountains of New York, an especially sensitive region, 30 percent of the lakes were receiving acid rain during 2006-08 in excess of the level needed to prevent harm." The report also highlighted the need for better information, *including the interactions of various pollutants.* (4) [italics mine]

As has earlier been noted, the amended Clean Air Act of 1990 also authorized the EPA to first establish a list of known hazardous air pollutants or HAPs and then to regulate them. HAPs are defined by the EPA as "those pollutants that cause or may cause cancer or other serious health effects, such as reproductive effects or birth defects, or adverse environmental and ecological effects." The original list contains 188 such HAPS, including pesticides such as captan, lindane, heptachlor, chlordane – and phosphorous, which is widely used in fertilizers as well as pesticides.

Apparently the EPA has a long way to go in its endeavor to protect the public because according to one watchdog group "[o]ver 5 billion pounds of toxic chemicals are released by industry into the nation's environment each year, including 75 million pounds of recognized carcinogens." (5)

As of December 7, 2009, as a result of a 2007 U.S. Supreme Court decision, CO2 was added to the EPA's short list of most common air pollutants. (6) While "ground level" ozone did appear on the original EPA list of six common air pollutants, neither CO2 nor ozone appears on the HAPs list. And although ground level ozone (unlike CO2) is not said to contribute to global warming, both are naturally occurring greenhouse gases, and both can be formed as a result of human activity.

In the case of ozone, we see that the EPA, like other sources, has somewhat curiously taken to using the terms photochemical smog (generally shortened to just plain smog) and ozone interchangeably. According to the EPA "Ozone (O3) is a gas composed of three oxygen atoms. It is not usually emitted directly into the air, but at ground-level is created by a chemical reaction between oxides of nitrogen (NOx) and volatile organic compounds (VOC) in the presence of sunlight. Ozone has the same chemical structure whether it occurs miles above the earth or at ground-level . . . Motor vehicle exhaust and industrial emissions, gasoline vapors, and chemical solvents as well as natural sources emit NOx and VOC that help form ozone. Ground-level ozone is the primary constituent of smog.. ." (7)

In other words, ground level ozone does not come out of tailpipes, smokestacks or chemical labs. It is a naturally occurring gas that is formed *in response* to the presence of primarily man-made chemicals, together with sunlight.

It is interesting to note that Webster's Seventh Collegiate Dictionary defines ozone as "pure and refreshing air." Another name for ozone is activated oxygen. In the atmosphere, ozone, or activated oxygen, is created when the oxygen released by plants drifts upward and is transformed into ozone by UV radiation. In fact, production of ozone depends on the availability of sunlight, which may help explain the hole in the ozone layer that appears over the poles in winter months when there is less sunshine.

Some atmospheric ozone continually drifts downward to lower levels of the atmosphere. Importantly, "[a] bit of ozone is needed in the lower layers of our atmosphere because it plays an important role in the chemical balance and degradation processes in our air." (8) Ozone is also created naturally at the earth's surface, that is ground level, by waterfalls, ocean surf, during thunderstorms and by lightning.

Interestingly enough, one recent paper published in the *Journal of Geophysical Research* found that "each flash of lightning on average in the several mid-latitude and subtropical thunderstorms studied turned 7 kilograms (15.4 pounds) of nitrogen into chemically reactive NOx. 'In other words, you could drive a new car across the United States more than 50 times and still produce less than half as much NOx as an average lightning flash,' [researcher] Ott estimated. . . . 'Lightning is one of the smaller factors for surface ozone levels, but in some cases a surge of ozone formed from lightning NOx could be enough to put a community out of compliance with EPA air quality standards during certain times of the year,' said [co-researcher] Pickering." (9)

What does one do with such information? Can we expect to find somewhere down the road a technological method to control lightning strikes? Or might we instead seriously address anthropogenic smog sources and explore the functions of ozone assigned by nature? Indeed, reference to ozone's natural function is part of the conclusion in at least one university level course outline for crop soils and oxidative stress in plants: "Activated forms of oxygen are important in the biosynthesis of "complex" organic molecules, in the polymerisation of cell wall constituents, in the detoxification of xenobiotic chemicals and in the defence against pathogens. . . " (10)

Surprisingly, activated oxygen, or ozone, has been widely used in medical applications for well over a century despite the fact that it was essentially criminalized in the US in 1976 when the FDA first announced in the Federal Register that "Ozone is a toxic gas with no known medical uses." (11)

Regardless of the FDA's seemingly unequivocal position, "at least 12 states have passed legislation to ensure that alternative therapies are available to consumers. Physicians in those states can legally use ozone as an alternative treatment in their practice, without fear of prosecution." In addition, medical ozone therapy is recognized in a number of foreign countries including France, Germany, Israel, Italy, the Czech Republic, Mexico, and Russia. (12) Certainly and at the very least, the relevant documentation for medical uses of ozone deserves our careful investigation. (13) (14)

The jury is also out as to whether or not ozone produces health problems, with one journalist asserting that "Researchers have been unable to kill animals even with long-term exposure to ozone at 10 times the levels found in the most polluted U.S. Cities." This same journalist maintains that "Ozone also can't be causing people to develop asthma. Asthma prevalence has nearly doubled during the past 25 years, but ozone and every other air pollutant sharply declined at the same time . . ." (15) While apparently comfortable about lumping ozone together with other air pollutants, this journalist is not alone in harboring doubts about ozone-induced asthma.

The reader may want to pause here to reflect that our own EPA has said that ozone does not come out of tail pipes or industrial plants, being formed instead *in response* to what *does* come out of those and other sources. This leads to the natural function assigned by nature to ozone. As we saw earlier, plant biologists know that ozone aides in the detoxification of xenobotic chemicals and in the defense of pathogens.

Thus, we find that the above journalist is not alone in his defense of ozone. And we can add to the list. For example, another researcher reports that a new study using data made available to the public by the N.C. Division of Air Quality and the State Center for Health Statistics shows there is, if anything, an inverse correlation to asthma: the higher the ozone level, the lower the asthma rate. (16)

This was confirmed by another report prepared by the New York State Department of Health for The U.S. Department of Health and Human Services Agency for Toxic Substances and Disease Registry: "This report compares ambient levels of certain hazardous air pollutants, criteria pollutants, and bioaerosols in two New York City neighborhoods that have different rates of hospital admissions for asthma and different socio-economic status characteristics. Chemical and biological analytes were chosen for this study based on existing information suggesting that exposure to these analytes may be related to acute asthma exacerbations. . . . . *Ozone tended to be negatively associated with all other analytes. . .*" (17) [italics mine]

When all is said and done, it does indeed appear that, despite claims that asthma and other upper respiratory diseases are caused by ozone, "[n]o one really knows the causes of asthma." (18)

Addressing the Real Culprits

It is, in sum, man-made, highly polluting chemicals, together with industrial-style agricultural and other land use practices (including climate remediation), that we should be primarily concerned about and not CO2 (or for that matter ozone). Indeed it is these kinds of negative externalities which economists wanted to account for when they first fashioned the cap and trade system. Instead, carbon pricing has shifted our focus away from pollution from all sources and caused us to become fixated on atmospheric carbon reduction, paving the way toward privatization of the atmosphere and the world's rain forests, while at the same time ignoring the crucial importance of carbon-rich, healthy soils in which carbon levels are, technologically speaking, impossible to measure.

What is needed is a redirection of our collective conscience towards the real culprits of environmental destruction, as well as how that destruction impacts climate, our soils and life itself. Part of our journey should of course include an examination of the major chemicals found in smog, which include sulphur dioxide, volatile organic compounds or VOCs and nitrogen oxides. (19)

For example, and according to the EPA, "[i]n the United States, about two-thirds of all sulfur dioxide and one-quarter of all nitrogen oxides come from electric power generation that relies on burning fossil fuels like coal." (20) As mentioned, both sulfur dioxide and nitrogen oxides are major components of acid rain.

Although produced naturally from active volcanoes and forest fires, the vast majority of sulphur dioxide is formed when coal or oil is burned as well as through other industrial processes. It "is toxic to a variety of plant life and may produce visible signs of injury and/or reduce yields of certain crops." (21)

In terms of human health, "[l]ong-term exposure to persistent levels of sulfur dioxide can cause chronic bronchitis, emphysema, and respiratory illness. It can also aggravate existing heart disease. . . Prolonged industrial exposure to sulfur dioxide may decrease fertility in men and women. . . Breathing sulfur dioxide can irritate the nose, throat, and lungs, and cause coughing and shortness of breath. Short-term exposure to sulfur dioxide can cause stomach pain, menstrual disorders, watery eyes, inhibition of thyroid function, loss of smell, headache, nausea, vomiting, fever, convulsions, and dizziness." (22)

Similarly, nitrogen dioxide is toxic by inhalation and is also considered a large scale pollutant. According to the EPA, "[n]itrogen dioxide (NO2) is one of a group of highly reactive gases known as 'oxides of nitrogen,' or 'nitrogen oxides (NOx).' . . . NO2 is the component of greatest interest and the indicator for the larger group of nitrogen oxides. NO2 forms quickly from emissions from cars, trucks and buses, power plants, and off-road equipment. . . . NO2 is linked with a number of adverse effects on the respiratory system." (23)

Notice that inhalation of CO2, even at levels used in greenhouses, will not produce the negative health effects of sulphur dioxide or nitrogen dioxide, nor are the high levels of CO2 used in greenhouses damaging to plants or the humans tending them – and greenhouse plants actually thrive on these high levels of CO2! Despite this, the

EPA in 2009 added CO_2 to its short list of common air pollutants – this decisions stemming from a 2007 Supreme Court decision.

Ozone, as mentioned, already was on the EPAs short list of common air pollutants. Interestingly, the EPA and other sources agree that ground-level ozone concentrations are often *low* in busy urban centers and *higher* in suburban and adjacent rural areas, particularly on sunny days in summer. This anomaly is explained away by the actions of wind and air. (24) Yet background ground level concentrations of nitrogen dioxide in some rural areas average around 30 µg/m3, not far below unhealthy levels. (25) No one as of yet seems to be asking if there is a connection between higher concentrations of ozone in suburban and adjacent rural areas and the nitrogen dioxide emanating from fossil-fuel driven farm equipment combined with the use of nitrogen fertilizers in those rural areas.

It may come as a surprise to many that nitrous oxide (N_2O) is, according to the EPA, a powerful greenhouse gas that has a global warming potential 296 to 310 times that of carbon dioxide (CO_2), this based on a formula devised by the IPCC. (26) By way of comparison, methane is said to have a global warming potential 72 times that of CO_2 averaged over 20 years, according to the IPCC's Third Assessment Report. A major source of anthropogenic N20 emissions is from chemical fertilizers.

Unlike methane, which has an atmospheric life of only about twelve years, both chlorofluorcarbons and nitrous oxides can remain in the atmosphere 100 years or more. Yet so much anthropogenic N_2O is entering the atmosphere that researchers are now positing that nitrous oxide (N_2O) stands to replace "anthropogenic halocarbons" as the major factor contributing to the destruction of the ozone layer. (27)

Nitrogen Fertilizers and CO2

Chemical fertilizers, especially nitrogen fertilizers, are a major source of nitrous oxide (N2O) emissions – which in turn represents one of the compounds involved in acidification of our waterways and oceans. These nitrogen fertilizers are not only a well known, major contaminant of waterways, but they consume enormous amounts of fossil fuels when produced. For example, the manufacture of nitrogen fertilizer alone represents an estimated 30% of a conventional farm's energy consumption.

And it seems that corn, which heads the list of crops cultivated worldwide, soaks up the most nitrogen fertilizer. This according to researchers at the University of Kentucky who calculated that the energy cost of nitrogen fertilizer production for use on corn amounts to a whopping 42% of the total cost of corn production. This compared to 29% for energy used in the drying of the corn and 7% attributable to the plowing and discing of the field. (28)

But it doesn't end there. All chemical fertilizers, but especially nitrogen fertilizers, alter soil viability and impair its ability to sequester or store atmospheric carbon or CO2. In other words, the soil's ability to function in it's natural capacity as a carbon reservoir and life-giver is being destroyed by modern agricultural practices, which has anthropogenic nitrogen oxides, petrochemical pesticides and soil-depleting, plantation-style agriculture stamped all over it.

Indeed, some researchers contend that decades of continuous cultivation and its attendant use of nitrogen fertilizers has reduced carbon stores in fields by as much as 50%. (29) The enormous implications of such large soil carbon losses begs another as yet unasked question: What, if anything, might this decline in soil carbon stores have to do with the worrisome decline in world oxygen supplies, climate change notwithstanding?

While it is a fairly well known fact that "compared to prehistoric times, the level of oxygen in the earth's atmosphere has declined by over a third and in polluted cities the decline may be more than 50%. . . with potentially serious implications for our health", no one as of yet points to industrial-style agriculture as a main culprit. Yet nitrogen fertilizers are not only limiting plant health and growth but nitrogen fertilizer run-off is, through creation of algae blooms, choking off the oxygen supply to aquatic life in affected waterways, and creating dead zones as a result. (30) This at the same time that rain forests, the lungs of the earth, are being razed to make way for carbon sink/biofuel plantations. Likewise, within the oxygen cycle, one wonders how surface level ozone a.k.a. activated oxygen fits into the big picture.

These questions aside, yet another landmark study conducted by four researchers at the University of Illinois openly challenges the popular notion that nitrogen fertilizers help soils sequester carbon in its title *"The Myth of Nitrogen Fertilization for Soil Carbon Sequestration."* According to research member C. W. Boast: "In numerous publications spanning more than 100 years and a wide variety of cropping and tillage practices we found consistent evidence of an organic carbon decline for fertilized soils throughout the world and including much of the Corn Belt besides Illinois."(31)

In a separate interview concerning the same study, University of Illinois team member Mulvaney points out that "the message we're delivering in our papers really is a rediscovery of a message that appeared in the '20s and '30s." (32) Like the rediscovered message of the research team in this study, it bears repeating here that organic soil carbon [SOC] decline, whether due to over-cultivation or chemical fertilizers or both, leads to degradation of air and soil quality and negatively affects crop nutrients. (33) Historical evidence proves that soil degradation in turn leads to climate change.

Fertilizers represent just one of the growing list of agricultural inputs that have only recently been added to the liability column of our planet's environmental balance sheet. Over the last one hundred plus years, these tech-heavy inputs have literally exploded. Today, thanks largely to the mis-named green revolution, they include an astonishing array of herbicides, pesticides, fungicides, GMO seeds (with documented negative health, environmental and economic consequences), plus cloned animals and the attendant use of antibiotics, larvicides and similar petrochemical based animal-aides. All of these are accompanied by a jaw-dropping array of pricey, resource-hungry farm equipment and assorted technologies.

Big Ag's Foods Create Local Deserts and Waterway Dead Zones

Increased reliance on irrigation has become another presumed input of necessity for industrial-style agriculture, dependent as it is on thirsty, soil-depleting commodity crops like corn, rice, wheat and cotton and coupled as it is with the economically-driven yearning to use arid climates to also grow similarly thirsty fruits and vegetables for global consumption.

Particularly since the so-called green revolution began, the irrigation required to support industrial style agriculture has been voraciously sucking entire lakes, rivers and aquifers dry (34) and has "led to saline build-up [which negatively affects plant growth] in about 30 million hectares of the world's 240 million hectares of irrigated land, according to the UN Food and Agriculture Organization (FAO)." (35)

Desertification is one devastating result. (36) This alone is a massive and growing problem. Experts, based on photos from NASA, estimate that more than half of the United States is desertifying, including the nation's fabled and well-irrigated "breadbasket." Even worse, these same maps show that about two thirds of the world is desertifying. The extent of desertification, and the maps themselves, clearly show that "over-grazing" is not the sole, or even the principal cause of this process. Moreover, irrigated crop lands only mask the process and tempt us to ignore the kinds of systemic changes that must occur if we are to survive as a planet and as a species.

Today, it is estimated that agricultural activities are responsible for somewhere between 70 and 87% of the total fresh water used globally each year, with irrigation responsible for the lion's share of agriculture's fresh water usage. (37) And unlike water use by residential or even industrial customers, of which about 90% eventually replenishes water sources, the water used for irrigation is only about one-half reusable. "The rest is lost by evaporation into the air, transpiration from plants, or is lost in transit, by a leaking pipe, for example." (38)

Freshwater sources around the world are becoming frighteningly limited not only because of irrigation practices but also because of serious levels of contamination by chemical fertilizers and pesticides. Worse still, both saline and fresh water systems are becoming appallingly compromised as a direct result of the circularity and waste created by modern agricultural activity.

For example, the lion's share of soil-carbon-depleting, chemical-hungry, thirsty corn goes for purposes other than providing healthy food for a hungry world. This includes use in biofuels, nutrient-void, chemically-laden corn oil, and ersatz manufactured (plastic) foods such as chips and high fructose corn syrup which are strongly tied to obesity and a variety of health problems. So it is that despite the seeming abundance of such products, truly health-giving food comprises a very minor portion of total agricultural production.

For example, about 80% of the corn produced in the U.S. goes to feed livestock. It forms the main component of a scientifically designed diet, which in the case of cattle "results in severe stomach ulcers and other health problems before they are slaughtered at 12 months of age." (39) Moreover, these unhealthy, miserable animals, conveniently exempted from animal welfare laws, are warehoused in crowded factory farms, otherwise known as CAFOs (or concentrated animal feeding operations) complete with giant manure lagoons.

Unlike their grass-fed counterparts, these factory-farmed animals have been held responsible for a significant portion of "anthropogenic" methane emissions, both through "eructating" and through their contribution to manure lagoons. (40) (41) But perhaps worse than methane emissions is the factory farm contribution to the creation of dead zones in waterways. Time and again these drug-and-pesticide laden, urine-and-feces filled waste lagoons have malfunctioned, spilling their contents dozens and often hundreds of miles downstream, killing thousands upon thousands of fish and creating significant and sometimes massive dead zones in their wake. (42)

Likewise, run-off of chemical fertilizers has, since 1985, left some 6,000 square miles of dead zone in the Louisiana and Mississippi Gulf. This has been attributed in no small part to nitrogen and phosphorus fertilizers used in the Midwestern farm belt, whose main crops are corn, soy and wheat. (43)

Although this concern was, for a time at least, effectively muted by the 2010 BP Gulf catastrophe, it must be remembered that oil is modern agriculture's most important source of inputs, including its use in the production of fertilizers and other chemical bandaids. Public outrage over spills such as these along with public pressure for alternatives energy sources have prompted oil companies like BP to begin positioning themselves through an interconnected web of strategic investments and legal and political maneuvers to take advantage of the potentially lucrative biofuels market as well as other forms of "clean energy" which all come with their own sets of environmental problems.

Large scale biofuel production, processing and use, for example, not only depends heavily on oil-dependent chemical bandaids, but also is connected with various forms of land, water and air pollution and actually results in a net *addition* of CO2 to the atmosphere. (44) Biofuels also contribute to hunger, especially among indigenous populations of the Global South where both land and food crops get diverted to the more lucrative biofuels market.

So it seems that in our erstwhile attempts to feed the world we have instead managed to wed the web of life to commodity agriculture and its new "green revolution" sidekick, all segments of which are reaping huge new profits from the climate change debate. The result is that 9+ million people worldwide face extinction through starvation each and every year and another 1+ *billion* suffer from chronic and unrelenting hunger. Yet another 2 billion suffer from micronutrient depletion. (45) (46) This in a world so awash with industrial-style food crops that surpluses are now being dumped into the new biofuel markets while tree plantation carbon sinks are daily created in delicate wetlands areas.

Without question, the unholy alliance between global warming and Big Ag (together with Big Energy and Biotech as its partners) is wreaking incalculable damage to planet earth and the web of life upon which we all must depend. It is an alliance we can ill afford to ignore and yet our default endorsement of the CO2 hypothesis, particularly as it relates to current carbon reduction methods, is the very thing allowing this alliance to grow in strength and influence.

3

RETHINKING CARBON SINKS AND SACRED COWS

Fortunately for us, and as testament to Mother Nature's resiliency, multiple studies have shown that "degraded soils can store up to five times more organic carbon in their surface layers than they currently hold [provided] soil management approaches change." (1) These kinds of soil management changes may even mitigate or reverse altogether whatever changes in regional and global climate that have been wrought by thousands of years of misguided land use practices. This alone could prove to be of significant impact. (2)

But what, exactly, might these soil management changes entail? Reducing cultivation requirements, clear cutting and vegetative denuding through chemical applications are important, obvious first steps. However, even these relatively obvious changes in soil management approaches are a much higher hurdle than appears on the surface, because underlying our tolerance of modern, industrial-style agriculture and its misnamed green revolution is the tragically mistaken notion that this is the only means by which to feed the world.

Yet it can be readily observed that said agriculture is failing miserably in this noble goal. Importantly, this failure to feed the world is not due to over-population or insufficient food crop but rather because the economics driving global geopolitics is forcing small farmers and self-sustaining indigenous peoples off the land even as it funnels potential food crops into biofuels and pushes delicate wetlands toward the development of carbon sinks and mono crop plantations.

All this at the same time that the legal matrix supporting agribusiness criminalizes seed-saving and legalizes pricey and profitable terminator seeds, one of many moves that is decimating the world's small farmers and the life-giving local foods their communities rely upon. (3) The tendency to overlook such factors skews our perceptions about over-population, malnutrition and world hunger and causes relevant statistics reported by well meaning activist groups to be misleading at best.

While it is true that world hunger is increasing, with some 1+ billion suffering chronic hunger and another 2 billion suffering from micronutrient malnutrition (4) (5), increasing hunger and outright starvation is in reality occurring at the local level, within select populations who have been deprived of traditional methods of growing food. Conversely, micronutrient malnutrition is a global problem and perhaps even more insidious because it is almost completely hidden from view. This micronutrient malnutrition is not just the result of chronic hunger; it is far more prevalently due to soil degradation taking place throughout the various stages of our modern, vertically and horizontally integrated food-and-fuel systems which are anchored by agribusiness.

All in a world so awash with food that the prices fetched often are below the cost of production, hence the need for subsidies. These subsidies almost wholly go to large scale commercial farming operations. Contrary to popular belief, most small and medium farm households in the U.S. are forced to rely on off-farm income, not subsidies, to help make ends meet. (6) Moreover, and in an expanding web of negative consequences, subsidies provide not only price and income protection primarily for large-scale farming operations, but these subsidies also provide additional support mechanisms, including marketing support, which help to ensure farmer use of high-priced technologies and toxic chemicals. (7)

Such subsidies and related "free" trade policies are an integral part of the legal matrix that contributes to hunger and malnutrition because they help force down prices the farmer is able to command to below the cost to produce. As mentioned, artificially low prices force small and medium farms in the U.S. to make ends meet with off-farm jobs, while those small farmers in poor countries often simply go hungry, and deeper into debt. This at the same time that soil-depleting mono cropping and intensive use of chemicals and all manner of hi-tech innovations are supported, even promoted, through subsidies and continually lower prices are forced through related "free" trade policies.

Biofuels, themselves subsidized on the assumption that they will wean the world off fossil fuels, also contribute to local hunger and pollution as both land and food crops get diverted to the more lucrative biofuels market. Finally, the rapid expansion of heavily subsidized agribusiness-related activities into emerging markets as well as third world countries via "free" trade is a growing problem, contributing to world hunger (among select populations), malnutrition (among the global population at large) and widespread environmental pollution and soil destruction.

One case in point is India where two-thirds of the nation still makes its living from the land. Sadly, this largest group of small farmers yet left in the world is facing imminent extinction. Why? Because agribusiness is "delinking farming from the earth and linking it to global markets and the corporations" that run them. As a result, Mother Earth, which had been "the most generous employer in this country of a billion, that has farmed this land for more than 5000 years" is under heavy assault emanating from heavily subsidized monocrop or plantation-style agriculture and its sidekick, the so-called green revolution. (8)

We need to be clear on this. Hunger and malnutrition are due to legalized poverty, deprivation of communal lands and the exceedingly reckless disregard for the looming specter of peak soil. Hunger and malnutrition are NOT due to over-population, or declining crop yields. Relevant data proves that the world is awash with food crops, appallingly nutrient-deficient though these may be, provided these crops be used as food. (9) This being said, it is, of course, abundantly clear that Planet Earth and the humans that depend on her cannot sustain much longer the unchecked cultivation of cardboard foods.

The Lessons of History

The problem is not one of simply feeding the world, but one of how best to nourish both the people of the world and the planet that supports us. As it turns out, anthropologists, archaeologists, food historians and others are finding that the lessons of the past can provide enormously valuable insight into the difficulties that have been historically associated with agriculture, and even its surprising relationship to climate change. In turn, a path to a brighter future might be constructed from these insights.

Montana rancher and award-winning journalist Richard Manning provides an inspired and historically accurate summary of the history of Western agriculture, as revealed by these various disciplines, that unveils some of the more important lessons and experiences of our ancestors:

> "[Mesopotamia] is the center of domestication for the Western world's main crops and livestock, ground zero of catastrophic agriculture.
>
> . . . .Wheat was the empire builder; its bare botanical facts dictated the motion and violence that we know as imperialism. . . .

. . . .Plato's lament is rooted in wheat agriculture, which depleted his country's soil and subsequently caused the series of declines that pushed centers of civilization to Rome, Turkey, and western Europe. By the fifth century, though, wheat's strategy of depleting and moving on ran up against the Atlantic Ocean. Fenced-in wheat agriculture is like rice agriculture. It balances its equations with famine. In the millennium between 500 and 1500, Britain suffered a major 'corrective' famine about every ten years; there were seventy-five in France during the same period. The incidence, however, dropped sharply when colonization brought an influx of new food to Europe. . .

The precolonial famines of Europe raised the question: What would happen when the planet's supply of arable land ran out? We have a clear answer. In about 1960 expansion hit its limits and the supply of unfarmed, arable lands came to an end. There was nothing left to plow. What happened was grain yields tripled.

The accepted term for this strange turn of events is the green revolution, though it would be more properly labeled the amber revolution, because it applied exclusively to grain—wheat, rice, and corn. Plant breeders tinkered with the architecture of these three grains so that they could be hypercharged with irrigation water and chemical fertilizers, especially nitrogen. This innovation meshed nicely with the increased "efficiency" of the industrialized factory-farm system. With the possible exception of the domestication of wheat, the green revolution is the worst thing that has ever happened to the planet.

> . . .More relevant here are the methods of the green revolution, which added orders of magnitude to the devastation. By mining the iron for tractors, drilling the new oil to fuel them and to make nitrogen fertilizers, and by taking the water that rain and rivers had meant for other lands, farming had extended its boundaries, its dominion, to lands that were not farmable. At the same time, it extended its boundaries across time, tapping fossil energy, stripping past assets." (10)

Exploitation of unfarmable lands may well prove to be the last frontier of the green revolution, and perhaps signal the end of life as we know it. Over two hundred years ago, America began to extend her land use boundaries to low-lying coastal wetlands in order to make way for port cities such as Boston and the nation's capital Washington, D. C. as well as to convert swamps into land destined for agricultural use. Indeed, for millennia before that, wetlands like these once were have been drained and denuded for agricultural use and for urbanization. In more modern times, they have been increasingly flooded for use as recreational lakes.

Too often unheeded over all these years were the elegant and intricate roles these wetlands play in the web of life. Lessons learned and then forgotten have taught us more than once that wetlands not only function as important flood control mechanisms but they also perform extremely important water filtering and cleaning functions, including purification of that which comes from the agricultural runoff of farms that replaced the wetlands in the first place. (11)

Climate Change: the Ultimate Price for Exploiting Unfarmable Lands?

The EPA has defined wetlands as those lands where water saturation is the main feature determining both the nature of soil development and the types of animals and plants living in and on the surface. They make note of the fact that wetlands will differ widely from one location to the next due to a variety of regional and local factors, including differences in soils, climate, water chemistry, topography, and vegetation, among other factors. And they say that wetlands are found "from the tundra to the tropics and on every continent except Antarctica.

Peatlands are an important category of wetlands and are distinguishable by the fact that they consist primarily of living, peat-forming plants. Technically speaking, peatlands include areas or regions that are dominated by peat to a depth of 30-40cm, even if the land has been drained. Peat itself can be thought of as partially carbonized vegetative matter, which has for centuries been used as a main source of fuel by indigenous populations. Considerable disputes exist over whether peat can be thought of as a renewable fuel, although the IPCC has begun to classify it as a "slow-regenerative" fuel. The burning capacity of peat has however become a major problem in emerging markets such as Indonesia where draining and logging activities are leading to widespread, difficult-to-control fires.

Peatlands comprise about 60% of the world's wetlands. The largest modern-day peatlands are in the boreal forests located in the cold tundra regions of Canada and Northern Russia. Interestingly, an ancient boreal forest existed in a region once called Mesopotamia. Also known as the fertile crescent, the cradle of civilization and the Garden of Eden, this region is what Richard Manning describes as ground zero of catastrophic agriculture. Today much of this fertile crescent is covered by desert and represents but one historical example of how man has altered and in this case warmed his climate through commodity style agriculture and clear-cutting. (12) (13)

25% of the planet's modern peatlands have likewise been destroyed and the peatlands of Europe and Central Asia have been severely impacted by mining and agriculture. Worldwide, agriculture accounts for about 50% of the damage done to peatlands, and forestry accounts for another 30%. According to Wetlands International, a science-based politically oriented advocacy group:

> "Efforts to improve the agricultural production of peatlands by conversion, drainage and fertilizing the soils are generally unsuccessful and very unsustainable. These areas often turn into wastelands within decades when as a result of soil subsidence they become prone to long periods of inundation. Many tropical peatlands are underlain by potential acid sulphate soils which become active when drainage levels become too deep, causing severe acidification of the area and its surroundings. Drainage also causes significant increase of fire risks." (14)

Southeast Asia contains the largest areas of tropical peatlands, and its tropical lowland peatlands were originally covered by dense rainforest. Interestingly and with the exception of Hawaii and Australia, all tropical rain forests can be found in developing and third world countries. And like India (not to mention pre-Colonial America), all of them "have indigenous populations who had their own system of land management and/or ownership in place for thousand of years before the intervention of colonists from rich industrialized nations." (15)

Rain forests have been described as the "lungs of the world" for the amount of oxygen they "exhale" and CO2 they "inhale." More generically, "[p]eatlands are the top long-term carbon store in the terrestrial biosphere and – next to oceanic deposits – Earth's second most important store." (16)

Peatlands also emit significant amounts of methane, which we are reassured is offset by the amount of carbon they sequester. Moreover, "[b]y sequestering and storing an enormous amount of atmospheric CO2 peatlands have had an increasing cooling effect, in the same way as in former geological eras, when they formed coal, lignite and other fossil fuels." Yet, because peatlands are often considered by geopolitical forces to be among "the last expanses of undeveloped land not in private ownership" they "are increasingly targeted by development that needs large areas of land, such as airports, plantations, wind farms and reservoirs." (17)

Indonesia for example is home to the most extensive rainforest cover in Asia, yet today the cover is less than 49 percent of what it originally was, and much of that is comprised of logged-over and degraded forest. Most often, it is oil-palm/carbon-sink tree plantations that are replacing natural forests, and much of the new lands being cleared for these plantations are cleared by burning. The burning of peatland in turn creates a whole new set of environmental and economic problems for local governments and the people. Through the entire process, wildlife is decimated, and the poor are dispossessed, sometimes forcefully and illegally. Uncounted millions are pushed to the outermost fringes of existence. (18)

The future may be even bleaker for the Amazon Rainforest, and thus the planet. This, the world's largest rainforest, has a land mass representing 40% of the total land in South America, extending into eight countries. Amazingly, the Amazon Rainforest produces more than 20 percent of the world's oxygen. (19)

As in Indonesia, the Amazon Rainforest has been undergoing massive transformations similar to but in some ways even more problematic than what is occurring in Indonesia. Unlike Indonesia there isn't even a pretense of law to protect the Amazon Rainforest.

What results, to an even greater degree than in Indonesia, is a very troubling collection of vested interests, private and public, local and international, who combine forces to enter into – legally, extra-legally and illegally – a variety of money-making schemes. These schemes include mining operations, various forms of industrial development, methane-emitting, earthquake-related hydroelectric dams and logging and large scale agriculture which continually takes over both undisturbed and logged rainforest areas as a means of providing land for commodity-style food crops, or for tree plantations or for cattle. (20)

Nikolas Kozloff, author of a book prophetically titled *No Rain in the Amazon* says that ranching and soy go hand in hand in this process, with soy planters buying up property cleared – and then discarded – by large scale ranching operations. These soy magnates also buy up transitional forests and savannahs, pushing slash-and-burn subsistence farmers and others farther into the forests.

One of the more haunting results of the soy takeover, says Kozloff, "is the so-called 'soy highway' . . . Over the past twenty years or so, soy plantations here have sprung up in the vicinity of the highway and now dominate the landscape. At times, you'd swear you were in Iowa or the American Midwest, and there's no wildlife to be seen for miles. At harvest time, fleets of green and yellow combines crisscross local fields lining the highway. John Deere dealerships are proliferating."(21)

As the soy highway has extended its tentacles across Brazil, elsewhere in South America public officials and experts have fallen prey to the so-called green agenda. For instance, "[f]orestry officials and foresters . . . are trying to convince public opinion that the country's cattle-related methane emissions are very high and that the country could 'compensate' [these emissions] by establishing carbon sink tree plantations. . . Surprising as it may seem, Argentinian and Uruguayan authorities seem to have forgotten that grassland soils are rich in organic matter, which means that they constitute huge carbon reservoirs." (22)

The Zimbabwe Struggle Is Emblematic, and Monumentally Important

Tree plantations and other carbon reducing projects undertaken by investors of the industrialized countries of the Global North along with those in newly emerging markets such as Brazil, Indonesia and China are not limited to the rainforest. They are extending to the dry or arid climates of emerging and third world regions as well, effectively obliterating local initiatives to rebuild the soil and depriving local populations of their means to a livelihood.

Indeed, one British humanitarian group known as Action Aid has said that more than 193,000 square miles have already been planted with biofuels plants around the world. Much of this activity has been taking place in Africa. In addition, there are eucalyptus plantations for production of pulp for paper and cellulose products, rubber plantations for rubber and numerous other monocrop plantations – all ostensibly established as carbon sinks, with added revenues coming from biofuels and assorted other agricultural products.

One devastating side effect in the creation of tree plantations, and all large scale cash crop operations, is that they wipe out up to 99% of the original wildlife, and that which remains is often dealt with in gruesome ways. As Montana rancher Richard Manning relates: "In rural Michigan, for example, the potato farmers have a peculiar tactic for dealing with the predations of whitetail deer. They gut-shoot them with small-bore rifles, in hopes the deer will limp off to the woods and die where they won't stink up the potato fields." (23) What the potato farmer understands in his corner of the world is that "unmanaged grazing" leads to resource degradation, which in his view is his potato crop.

Another equally devastating side effect of tree plantations and other forms of industrial or large scale agriculture is that the soil is destroyed and the rivers become silted over, as happened in Mesopotamia thousands of years ago. Such is the case in modern Zimbabwe, which was once known as the breadbasket of Southern Africa due to its wheat production. Since 1980, Zimbabwe has been moving away from cash crops such as wheat, sugar cane, tobacco and cotton, and into cattle rearing.

Today some 80% of Zimbabwe is rangeland of some kind. However, "just about all of it is degrading, contributing to the fact that Zimbabwe doesn't have any more internal rivers." Why? Because, not unlike old Mesopotamia, "they are all silted up. They run briefly, after heavy storms, with muddy water. What was once a web of real rivers (much of Zimbabwe is a watershed for Mozambique) is now sandy river beds. More and more water runs off to Mozambique, sometimes causing severe flooding in parts of that country."

Yet in Dimbangombe, which is one of the more devastated regions of Zimbabwe, "cattle rearers" are showing how one can restore the land's health and thus the water through the way one manages livestock. Here, "[h]erders are successfully using cattle to restore the land and to regenerate the rivers . . . They are demonstrating what was once known but has been widely forgotten: that cattle and other large herbivores play a vital role in maintaining ecosystems in arid parts of the world. They are working with nature, not against it." (24)

A potential boost for these herders came in the form of a study reported in *Arid Lands Newlsetter* of the University of Arizona, which concluded that marginal agricultural areas converted to rangeland should be included as accountable carbon sinks. (25) Unfortunately for local Zimbabwe ranchers seeking to provide a reliable, nutrient dense local food source, investors have, despite the recent speed bump in the carbon markets, been looking to accrue carbon credits through implementation of tree plantation projects, resulting in what amounts to a giant land grab which is hardly unique to Zimbabwe. (26) (27)

The evolution of these carbon-reduction schemes is endless. For example, one of the latest proposals to help sequester "excess" CO2 is to create grassland sinks in place of that portion of cultivated, CO2-emitting land which is now used to produce "surplus" food. Since removal of excess CO2 through development of grassland carbon sinks is the main goal, long standing worldwide problems with hunger and starvation get pushed aside, as does the health and nutrient value of food crops grown on carbon depleted soils.

Likewise overlooked, as in the case of the potato farmer, is that "grasslands and grazers have co-evolved over millions of years, [which means that] grasslands NEED grazers, be they kangaroos, elephants, termites or sheep, to facilitate energy flow and the recycling of nutrients. [Particularly in] medium to low rainfall areas, grasses which are not grazed become senescent and cease to grow productively. . . " Furthermore "[m]anaged grazing is arguably the only natural process by which grasslands can be 'improved' on a sustainable basis. Unmanaged grazing, or complete exclusion from grazing, will inexorably . . . lead to desertification in all but the high rainfall areas."(28) Just as important, grazed pastureland is proving to be the *best* land use for storing carbon, so much needed in today's depleted soils! (29)

Critically important to the management of grasslands, say journalists Allan Savory and Christopher Peck in an article appearing in the 2010 summer issue of *Green Money Journal*, is not only to understand that "large numbers of grazing animals are essential to the health of the grasslands, but they must be in large herds constantly moving as they once were when accompanied by pack-hunting predators. . . Practice and experience around the globe has demonstrated that mimicking nature in this way works: rangelands regenerate, the diversity of plant species and their numbers increase, and there is less soil erosion, fewer droughts. A highly beneficial consequence of managing animals in this way is the increase of soil micro-fauna. . . " (30)

It is worth noting that farm and grazing lands managed this closely can only be managed personally by small operators, not long distance by suitcase farmers (31) which include investors, financial institutions and multinational conglomerates such as ADM or Conagra, all of whom must bow to the short term demand of extra levels of profit-taking as well as interest payments on their debt. Moreover this small operations model has benefits far beyond the merely economic, and this is especially true when the food produced is only sold locally, that is to say within a 200 to 300 mile radius.

Nourishing the Soil and the People at the Same Time

Virginia-based celebrity farmer Joel Salatin is one small operator pioneering in intensive grazing management, with a policy of selling within a 150 mile radius. His farm, called Polyface Inc., is a family operation that has been featured in *Smithsonian Magazine, National Geographic, Gourmet* and a long list of other radio, television and print media.

Through Salatin's version of intensive management, which he calls perennial polyculture, Salatin has achieved the previously unimaginable by *adding* carbon rich material and humus every year to his farm all while producing prodigious amounts of nutrient-dense food in the process. All is accomplished on site and without Big Ag's help, via land management techniques which are carefully designed to imitate the structure of a natural system whose needs, like those of peat and wetland biomes, are unique to his specific farm.

Comprising roughly 550 acres, the farm portion is about 100 acres of open land and the balance is wooded. Altogether Polyface "supports a mixed bag of free-ranging animals. Annual production amounts to about 900 head of cattle, 30,000 broiler chickens, 4,000 layers, 800 pigs, [plus about 1,000 turkeys], and 1,000 rabbits, plus 450 acres of timber [which is sustainably harvested]." (32)

Salatin says he currently feeds between 7,000 and 9,000 locals and asserts that intensive grazing management is "the ONLY system that really can feed the world." By way of partial explanation he says: "Let's take cows, for example. With good, controlled grazing, where we're moving these cows every day to a new patch, on our farm we're averaging 400 cow days per acre – a cow day is what one cow will eat in a day. Our county average is 80." (33)

Salatin is in other words, achieving five times the "cow days per acre" that conventionally managed grazing permits, and he does this without any grain being fed to his cattle! A key feature is that the SAME acreage used by cows is ALSO shared in symbiotic fashion by poultry and other livestock which are each rotated on and off the same acre according to the life cycle of specific and unique contributions of each animal, hence the term "intensive grazing management."

In addition to no grain being fed to his cattle, the entire process is accomplished with no fertilizers, or other chemicals. This results in a dramatic reduction of all inputs – while at the same time it is *adding* precious humus via a healthy mix of animal manures, native grasses and insects. And, because the food is only sold locally and not to international markets, the pollution footprint in transportation and storage is substantially reduced. Not only is Salatin able to earn a handsome six figure income, while also employing a small staff, but he gets to keep more of his income for himself due to his low overhead cost for fertilizers and other inputs.

Out of curiosity, let's test Salatin's assertion that intensive grazing is the only way to feed the world. We know that there are roughly 37 million square miles of Earth's terrestrial surface under cultivation, with about 24 million of that used for grazing and 13 million for what are loosely termed crops. (34)

Let's apply Salatin's favorite 80/20 rule (which means that he uses 80% of his land for forestry and 20% of his land for grazing) to world figures. The total world use of land for grazing is roughly 24 million square miles. 20% of that 24 million square miles is 4.8 million square miles which, per the 80/20 rule, will now be used for intensive grazing activities, which figure we will apply as a means by which to extrapolate what might be possible under a system similar to Salatin's. Again applying Salatin's 80/20 rule, the remaining 80% of the 24 million square miles now devoted to grazing activities, or 19.2 million square miles, will be hypothetically reserved for complex forest systems (as opposed to chemically denuded tree plantations), some of which will be sustainably harvested.

By multiplying 4.8 million square miles times 640 (the number of acres in a square mile) we get 3,072,000,000 intensive grazing acres, Polyface-style. Using Salatin's low estimate of 7,000 people that his farm feeds per 100 intensive grazing acres per year we find that we have an eye-popping total of 215 BILLION people that can be fed high quality, nutrient dense, low input food per year. In this manner and contrary to critics' estimates (35), the amount of broiler chickens alone that could be produced by the Polyface method works out to roughly 921 BILLION, per year. Consumers in the U.S. currently purchase less than 10 billion of these – and most, in direct contrast to the Polyface model, are factory farmed. (36)

But all this is not the end of potential earth-saving, earth-sustaining food production. Since more than 50% of U.S. grain and nearly 40% of world grain is fed to livestock, roughly 6 million square miles of the 13 million square miles now devoted to high input, industrial style grain crops can be retired and/or rehabilitated.

For example, and dependent on the local soil system, some of the freed-up acreage might be returned to multi-species perennial grasses and crop systems; some freed-up acreage to high quality, low input, vegetable gardens and orchards; and some freed-up acreage to properly rotated, locally grown grain to supplement chicken foraging and perhaps some winter hay for the cattle. The remaining acreage might be devoted to the production of a variety of local, organically grown grains and legumes to use for local artisan breads and similar locally produced foods. All is of course dependent on the idea that no new acreage be devoted to large scale biofuel production or other forms of plantation style agriculture, and that whatever acreage now devoted to such agriculture not be allowed to displace foods grown locally – by and for poor populations.

The Salatin model is but one example of a perennial farming system, which includes perennial crop systems, perennial-annual integration strategies like agroforestry, and, as in the Salatin example, livestock-perennial combinations. An added benefit, as research shows, is that this type of cultivation will result in increased rates of carbon sequestration, which could – most importantly – contribute to organic soil carbon stores, while building all important humus. (37)

These perennial farming systems are all based on strategies which improve the productive capacity of the soil over time, and as such can be thought of as regenerative agriculture, or if you like, carbon-sequestering agriculture. (38) Such systems, based as they are on small producers and local artisans, would also help immensely to revitalize the air, soil and water quality and the economies of now devastated rural communities everywhere.

The reader will recall from Chapter 1 that various experts have determined that certain forms of what we will now more accurately refer to as regenerative agriculture had the best potential for reducing GHG's, but discarded that solution in favor of carbon pricing as an easier if less effective approach to managing CO2 and by inference, climate change. Thus it remains to be seen whether perennial farming systems will ever come close to being achieved on a wide spread scale.

While growing interest, independent research, experience and need most certainly support such forms of regenerative agriculture, there are – as the experts who turned to carbon pricing knew – mountains of economic, social, political and legal hurdles to jump over first. All of it despite continual warnings that "world food production is at risk from farming methods that have degraded soils, parched aquifers, polluted waters, and caused the loss of animal and plant species." (39)

Arguably we still have time to reverse the current situation, but will we? More to the point, one has to ask whether it is even possible within the context of the current monetary system and the legal structure which supports it.

4

CLEAN ENERGY AND PEAK SOIL: QUI BONO?

In February of 2003, an article by award-winning journalist Richard Manning was published in *Harper's Magazine.* Almost immediately it gained a wide readership and is now available at numerous locations around the internet, including university web pages as part of course discussion. Its catchy and descriptive title, "The Oil We Eat: Following the Food Chain Back to Iraq" nicely captures his main thesis – which is that Big Ag is not about food. "It's about commodities that require the outlay of still more energy to become food." And so it is. Not only is said agriculture spreading rapidly around the globe but a careful understanding of it helps explain why Big Oil and Big Ag are so reliant upon each other for their existence.

Roughly around the same time that Manning's article made its debut, two of the world's largest oil giants, BP and Royal Dutch Shell, incongruously declared that by the year 2050 one-third of the world's energy will need to come from solar, wind and other renewable sources. The stage was set for Big Oil's foray into renewables, disingenuously regrouped under the banner of clean energy which now includes clean coal and clean gas – and as some would have it, even nuclear.

The reasons the oil giants gave are familiar ones: climate change, population growth, and fossil fuel depletion, all of which are understandable concerns, particularly within the context of Big

Energy/Big Ag business models. While excitement over cap and trade may have temporarily hit a speed bump, interest in renewable energy certainly has not. At the same time, a significant number of agricultural projects begun under cap and trade and now owned by Big Energy, as well as Big Ag, continue to provide revenue through logging, commodity crop production – and biofuel production.

In 2008, BP and Shell unceremoniously pulled the plug on their British-based renewable energy projects and moved them to the U.S. This despite prime minister Gordon Brown's promise "to sweep away all impediments" as part of a desperate bid to keep jobs from fleeing to the U.S. As in all things, money talks and it so happened that the U.S. was offering hard-to-resist clean energy incentives that also provided a convenient tax shelter for oil and gas revenues. (1)

This score for the U.S. is not only likely to be temporary, but costly, both in economic and environmental terms. For example, concerns over global warming have convinced most Americans, including their elected leaders, to more or less buy into the hype surrounding renewables aka clean energy, all of which come with some significant if unacknowledged environmental consequences, more about which later.

So palpable are present day climate change concerns that next-to-no mention is made, even in the midst of today's debt crisis, about the huge economic concessions (some would say bribes) given to Big Energy and an endless parade of start-up renewable energy related companies, nearly all of which are either subsidiaries of Big Energy or will be swallowed up by them.

The ultimate goal is clean energy, obtained through innovations that could either lead to cleaner burning fossil fuels or more reliable forms of renewable energy. The terms "clean energy" and "cleaner burning fuel" almost always refer to calculated CO_2 emissions. In the name of cleaner fuels, whatever they may be, untold billions are thus awarded on the optimistic if questionable assumption that U.S. energy independence, together with clean, green, reliable, big-grid energy, will be ours once we work out a long and growing list of technological kinks.

The most familiar and sanctioned of the renewables are biofuels, hydropower, geothermal, solar and wind. Substantial numbers inside both government and industry will also include the newly rechristened clean coal or even so-called clean natural gas and zero-carbon emission nuclear as part of the clean energy sector.

Yet the truth is that virtually all of these widely touted clean energy choices are either partially or wholly connected to a whole panoply of egregious land abuse practices ranging from legal, extra-legal and blatantly illegal land grabs, to the clear-cutting of rain forests, to mountain top removal, to toxic residue from mining for materials such as silicon for solar panels or coal for cheap electricity or uranium for nuclear power, to complete decimation of fresh water supplies, rain forests and peatlands, wildlife and local populations.

All of this in order to establish giant biofuel/agrofuel plantations or windmill "farms" or hydroelectric dams and more as the unsubstantiated means by which we are to supply the world's energy needs, in a "clean," green and cheap manner. (2) (3) (4) (5) (6) All of these energy alternatives are promoted on the basis of which energy source contributes the least amount of CO2 to the atmosphere, while tar sands and other dirtier forms of energy are promoted on the basis of consumer demand for reliable energy and the economic need to keep the conduits of commerce open. Environmental pollution and other forms of land and people abuse somehow get lost in the shuffle, even as the war over energy choices continues to escalate. Meanwhile, we find ourselves confronted with one energy disaster story after another.

Trouble in Techtopia

For example and while the newly rechristened clean coal industry together with our government doggedly await yet-to-be-developed technologies to give substance to coal's hopeful new clean image, seismologists have been uncovering some troubling connections between earthquakes and green technologies such as geothermal, hydropower and new carbon sequestration technologies, not to mention coal mining itself. The new concern is not so much that these projects will be damaged *by* earthquakes, wreaking havoc for those in their path, but that they can actually *cause* earthquakes thus "raising questions about the wisdom of energy projects and in some cases stopping them in their tracks." (7) (8) (9)

Earthquakes are just one of the problems associated with these so-called green technologies. A case in point is carbon sequestration technology, which is an integral part of coal's new clean image. It turns out that this technology "has a huge downside: It could nearly double the amount of water a [coal] plant uses for every kilowatt of electricity it delivers—easily erasing any gains from techniques aimed at conserving water." (10) Considering that Big Ag itself uses 70%+ of the world's fresh water supply every year, this type of water use is an even bigger problem than appears on first blush. (11)

Somewhat thankfully, growing public opposition to coal plants is not limited to CO2 emissions so the industry is facing an increasingly uphill battle, despite its proposed new clean image. Surprisingly, and despite the virtual tsunami of money being funneled into agrofuels and other so-called renewables, coal's primary competition is coming from the natural gas industry. Why? Because compared to coal, natural gas is seen as a cleaner burning fuel, and it is more reliable than, say, solar or wind which are subject to the vagaries of the weather.

To be sure, the highly controversial shale gas, made newly available through hydraulic fracturing (or fracking for short), has actually increased U.S. natural gas reserves, allowing prices to fall and enabling companies to contemplate its export. Simple stats tell the story of a boom that has managed to transform America's entire energy landscape almost overnight, because while "as recently as 2000, shale gas was 1% of America's gas supplies; today it is 25%." (12) Moreover, "recent world inventories indicate there is at least as much recoverable shale gas as there is conventional natural gas already inventoried. In response, natural gas prices have dropped close to 50 percent." (13)

While it's increasing affordability may help increase public acceptance, shale gas isn't in any way a clean, land friendly energy source. This assertion is made crystal clear in a May 2010 article titled *"The Dirty Truth Behind Clean Natural Gas,"* released by the National Wildlife Federation, as follows:

> "Energy sprawl is creeping across public lands in Wyoming, Montana, Colorado, New Mexico, Utah and the Dakotas. Nearly 120,000 wells (mostly natural gas) were drilled West-wide during the past decade, and 44 million acres of federal public lands are now leased to energy companies out of 256 million acres overseen by BLM. It is an energy boom bigger than any in the history of the West, and it is scarring some of the nation's last great wide-open spaces. Wherever gas is drilled, wildlife and its habitat are suffering. Gas fields are spreading noxious weeds, polluting the air and fouling the water. Public lands valued for grazing, hunting, wildlife-watching, recreation and other multiple uses are being converted to single-use industrial zones.

> Today, most of the remaining domestic reserves are so-called "unconventional" deposits trapped in shale, coal and sandstone formations. To free the gas, companies pump chemicals, sand and water into the ground under high pressure to fracture the rock formations, a process called fracking. Hydraulic fracturing fluids contain a toxic cocktail of petroleum distillates—benzene, toluene, and other carcinogens (the precise recipe is a trade secret). The fractured formations are then dewatered to release the gas." (14)

Fracking not only uses prodigious amounts of fresh water, but researchers such as Theo Coburn, together with residents living near such operations, steadfastly insist that it negatively affects water and air quality. Even the EPA reversed the findings of its own 2004 study when in 2009 it found contaminants related to those used in fracking operations in the well water of nearby residents in a small rural community in central Wyoming. (15) (16) (17) (18) (19)

Predictably, the industries and involved investors have rationalized away any negative externalities, and are now waging huge PR campaigns in their defense. Be that as it may, denial is becoming more difficult as evidence mounts. But even in the absence of evidence, one must seriously question why the industry saw fit to seek exemptions from the Safe Drinking Water Act in 2005, along with exemptions from parts of at least six other statutes including the Clean Air Act and the Clean Water Act.

Fracking, like carbon sequestration, hydropower and similar technologies, is also being connected to man-made earthquakes. As early as March of 2011 experts were warning that a "sudden swarm of earthquakes in Arkansas – including the largest quake to hit the state in 35 years – is very possibly an after effect of natural-gas drilling." (20) A few months later, fracking was actually halted in England "after scientists said two small earthquakes might be linked to [it.]"

(21) By April of 2012, two more studies documented direct connections between earthquakes and both oil and gas extraction and waste-water injection." (22)

Although some of these earthquake-inducing, soil-destroying projects may be, temporarily at least, put on hold, large institutional investors and the companies and governments they support all have powerful economic incentives, as well as intense, consumer-driven demand for energy, in seeing that these kinds of projects continue. While the various technological kinks are presumably being worked out, said investors can hedge their bets by turning to a veritable smorgasbord of other potentially big money-makers in the clean energy sector, not to mention tar sands and other forms of dirty energy.

Investment choices are of course based on a number of factors, including prevailing public sentiment, the type of government support available, and potentials for profit. Perhaps most important, however, is that these investment choices are for all intents propelled forward by the rules of a zero-sum game, in which the gains of one participant are achieved through the equivalent losses of another participant, resulting in a transfer of wealth from losers to winners. Especially since the industrial age, virtually all losses accruing from externalities such as land degradation, pollution and other forms of land abuse have been absorbed by the losing participant, who by and large has always been the local community and its producers.

The increasingly heavy burden of negative externalities is today compounded by the CO2 hypothesis, which itself encourages the transformation of CO2 into a tradable commodity. The new economic reality created by the carbon markets and the quest for renewables is foisting ever more of the costs of negative externalities onto local communities even as it facilitates the privatization of the atmosphere and the world's rain forests. It is on this basis that CO2 has become the axis around which all such potentially lucrative investments turn – land use or abuse thereof notwithstanding.

The CO2 Hypothesis Bows to Business As Usual, Ignores Peak Soil

Like all previous nuclear accidents, the Fukushima disaster of March 2011 initially caused public support for nuclear power to plummet. Yet despite the magnitude and on-going problems of Fukushima, nuclear power advocates remain firmly convinced that nuclear power was, and is, here to stay. To them, economic benefits, coupled with purportedly low environmental impact and zero greenhouse gas emissions are key reasons why the benefits of nuclear energy outweighs the risks.

As the Nuclear Energy Institute puts it, "nuclear energy is America's largest source of clean-air, carbon-free electricity." Various versions of this oft-repeated statement are based on the fact that nuclear power plants themselves emit no "criteria" pollutants and no greenhouse gases while generating electricity. Proponents such as the Nuclear Energy Institute also point out that "the life-cycle emissions from nuclear energy are comparable to other non-emitting sources of electricity like wind, solar and hydropower." These reasons together with the fact that nuclear energy production requires a relatively small amount of land (excluding land needed for "waste isolation") means that nuclear energy has one of the lowest impacts on the environment of any energy source. All of this is accomplished at the same time that nuclear power plants provide a reliable and significant portion of the nation's electricity needs at the lowest cost. (23)

Meanwhile, nuclear power opponents look to the entire production chain of nuclear energy, and strenuously object to the above green characterization of nuclear power. In so doing they rightly assert that "nuclear power is actually a chain of highly energy-intensive industrial processes. These include uranium mining, conversion, enrichment and fabrication of nuclear fuel; construction and deconstruction of the massive nuclear facility structures; and the disposition of high-level nuclear waste." (24) Sustainability is another problem since uranium, which is the energy source for nuclear energy, is a scarce resource, with supplies estimated to last less than 60 more years. These objections have remained substantially the same for many years, unlike the on-going refinement of "nuclear greening" campaigns waged by the Nuclear Energy Institute and allies. Fukushima has been, and still is, considered to be little more than a speed bump for nuclear power advocates even if it did provide some additional fuel for the on-going propaganda war.

Immediately following Fukushima, nuclear experts and industry officials declared the fear-mongering propaganda by the anti-nuclear "clubs" to be "excessive." They unanimously agreed that "the battle for the hearts and minds of the public must be won" specifically because nuclear energy is the only realistic replacement for fossil fuels that does not also emit greenhouse gases and because it reliably provides electricity at a low cost to a wide consumer base. Once the public comes to its senses and the fear-mongering subsides these experts predict that the long term viability of the nuclear industry would remain intact and would likely return to "a business as usual scenario after a slight hiatus for evaluation of both requirements and safety." (25)

Industry money has played an important role in effectively marginalizing the anti-nuclear "clubs" and securing both public and private support for the nuclear industry. For example, in March of 2012, the Nuclear Energy Institute issued a release in which it said that the nuclear industry was investing in record funding for education. At the same time it complained that Federal support was withering. Yet the Nuclear Regulatory Agency manages a page, last updated December of 2012, detailing an impressive list of grants and scholarships it had available for furthering nuclear education. And a September 21, 2011 article on Yahoo said that "the Department of Energy announced today that it is awarding more than $17 million in grants to 23 universities to strengthen and encourage the development of new nuclear energy technologies." This was in addition to the August announcement by the DOE that it would provide $39 million in nuclear research grants to 31 academic institutions in 20 states.

However, being outgunned by funding sources and attendant PR campaigns are not the only problems faced by nuclear power opponents because a good case can be made that they themselves invite marginalization when they make the premature case that renewables – including biofuels, wind and solar – are ready for large-scale use. They promote renewables by emphasizing, just as their nuclear advocate counterparts do, potential for CO2 reduction and/or energy independence without giving adequate attention to the ultimate environmental consequences of these various and sundry energy projects.

For example, it has been claimed that photovoltaic power from solar panels "has advantages over wind power, hydropower, and solar thermal power, [since] the latter three require turbines with moving parts that are noisy and require maintenance." (26) The truth is however that photovoltaic cells are not only inefficient, unreliable and extraordinarily expensive, but they use extremely rare materials and cause environmental contamination through both the mining and disposal processes. (27)

Similarly, recent studies have found that large windmills have a very high risk of contaminating aquifers and ground water during both the construction and life of the windmill farm. (28) (29) Maintenance and reliability are additional problems. For example, "in high winds, ironically, the turbines must be stopped because they are easily damaged. Build-up of dead bugs has been shown to halve the maximum power generated by a wind turbine, reducing the average power generated by 25% and more. Build-up of salt on off-shore turbine blades similarly has been shown to reduce the power generated by 20%-30%." (30)

As it happens, The Nature Conservancy published a report in August 2009 entitled *"Energy Sprawl or Energy Efficiency: Climate Policy Impacts on Natural Habitat for the United States of America."* The objective was to examine the extent to which climate policy that is formulated so as to reduce CO2 emissions ultimately results in what the authors called "energy sprawl," which is the amount of land that is required to produce energy from different sources. (31)

The findings of this report were not what tree-huggers want to hear, even as they provided ammunition for the zero GHG/CO2 emitting nuclear industry. As one dedicated hunting enthusiast summarized: "The report determined that nuclear energy is the least land-intensive, as it requires only 1 square mile to produce 1 million megawatt-hours per year, enough electricity for about 90,000 homes. To produce the same amount of energy, the report determined that geothermal energy requires 3 square miles, coal requires 4 square miles, natural gas needs 8, petroleum needs 18 and biofuels (corn and others plants) require up to 500 square miles. Wind farms require more than 30 square miles. [Solar requires over twice the land that wind does.]" (32)

Startling as these findings may be, the report did also point out that "many energy production techniques actually have multiple effects on biodiversity" beyond that of simply how much land these energy systems take up. These effects, or impacts, rightfully should be taken into account when considering various energy production techniques. They include acid rain and other forms of air pollution, eutrophication and other effects of water pollution, water consumption, diversion of natural water flows through hydropower dams, and toxic residue from mining. In other words, negative externalities.

All of this notwithstanding, it is still wind, solar and biofuels which require land in amounts that are orders of magnitude greater than nuclear, coal, gas and oil. And just as importantly, each of these renewables themselves create significant environmental problems that can be partially or wholly as damaging as that created by their non-renewable counterparts. So it is that, without listening and without hearing, opposing sides wage an endless war of words over whether renewables can ever replace fossil fuels (33), or whether "clean" coal, gas and nuclear can ever truly be both clean and cheap.

Meanwhile oil remains the most efficient source of energy, based on amount of land required. Whatever its environmental impacts may be can be justified within the context of the zero-sum game, where units of energy output are weighed against units of environmental damage. Oil wins hands down because once we compare joules to BTUs we find that in order to "obtain in one year the amount of energy contained in one cubic mile of oil, each year for 50 years we would need to have produced three Gorges dams, 52 nuclear power plants, 104 coal-fired plants, 32,850 wind turbines, or 91,250,000 solar panels." (34)

The inherent energy capacity of oil is certain to keep oil as a major world energy source well into the future, even as it's extraction, processing and transportation gets dirtier and more problematic over time. Why? Because it seems that Peak Oil has been put off into the relatively distant future due to the massive amounts of oil available in unproven reserves, including tar sands, shale oil and heavy oil now being made more economically feasible as a result of climbing oil prices.

This is affirmed by a 1999 Greenpeace report on Climate Change, which states that "a modest 3,000 billion barrels reserve of non-conventional oil . . . could then sustain a continued increase in world oil use beyond the middle of the 21st century on the basis of an assumption of a 2% per annum growth in demand. Under this scenario the world oil industry in the mid-21st century would be approximately three times its present size." (35)

Greenpeace is no longer the lone voice in this wilderness. The most recent estimates of gas and oil reserves provided by the USGS for the years 2009 and 2010 have shown spectacular increases in both U.S. gas and oil reserves, even leading one left-leaning journalist to proclaim that "the energy abundance that helped propel the United States to global leadership in the 19th and 20th centuries is back. . ." (36)

The Biofuels Non-solution

As the climate change/clean energy battles rage on, the topic that gets left out of virtually all discussions is that which has the most potential for pointing us toward the most promising energy solutions while at the same time providing systemic solutions to regeneration of the land – and yes, even climate change. That topic is land use, or abuse thereof. This void in the climate change/clean energy debate has become especially pronounced ever since the public's focus somehow shifted from real pollutants such as sulphur dioxide, nitrogen oxide, carbon monoxide and air particulates to greenhouse gases as being the primary culprits in climate change.

Large scale biofuels, or more accurately agrofuels, are a case in point. Produced from living organisms which run the gamut from soybean and rapeseed oil, palm seed oil, sugar cane and sugar beet, corn products, algae and more, these biofuels have been promoted as a green alternative to oil and gas. Agribusiness farmers were thus given a new reason to maintain plantation style agricultural practices, and mammoth processing plants began to be built across the United States and the world. Few if any thought to consider what the production, processing or burning of these biofuels might do to the land, the environment, or the planet's inhabitants.

As pointed out in the Nature Conservancy report discussed above, such biofuel plantations require inordinate amounts of land. In addition, their production virtually ignores the carbon needs and carbon sequestering effects of the soil even as applications of pesticides and other chemicals contaminate the soil and waterways thereby damaging, if not destroying the ecosystem. These important facts seem to escape all notice, even though biofuels themselves have not been without their critics. For example, one of the complaints now most commonly lodged against such fuels is that nearly "all biofuels used today cause more greenhouse gas emissions than conventional fuels if the full emissions costs of producing these 'green' fuels are taken into account." (37)

The truth is that these fuels do indeed result in increased greenhouse gas emissions – primarily in the form of nitrous oxides from fertilizers and methane from clear cutting of rain forests – and not CO2. But that is not the end of it, because both the production *and* processing of biofuels also result in truly appalling levels of soil, air and water contamination and a gross misuse of land and water resources. (38) (39) Moreover, the inordinate amount of land that is required to turn agricultural products such as corn or soybean oil into fuels (as pointed out by The Nature Conservancy report discussed above) almost pales in comparison to the damage that the production, processing and burning of these fuels do to the environment, not to mention the harm caused to indigenous populations and developing economies.

As agrofuel proponents would have it, these "green" fuels will increase national security by decreasing dependence on foreign oil. (40) Never mind that said green fuels are virtually always produced by heavily subsidized, vertically-and-horizontally-integrated multinational operations that depend on land grabs of formerly communal lands and clear cutting to make way for plantation-style "farms" heavily dependent on fertilizer and pesticide use. Never mind that said biofuels produce significant toxic residue from processing, and emit CO2 when burned. Never mind that said biofuels divert food production away from hungry populations or that they use inordinate amounts of fresh water in order to produce. (41) (42) All of it propelled inexorably forward by the great global casino, otherwise known as the international financial market – which now includes the carbon market.

By forging strategic alliances with biotech, petrochemicals and oil, large agribusiness corporations are, to a very large degree, able to dominate both the financial markets (and the profits derived therefrom) and the global energy agenda. This coalition of seemingly odd bedfellows then creates, through incessant propaganda, a web of self-reinforcing beliefs that help to further justify the biofuels "green" fairy tale. In this way, agrofuel proponents manage to convince themselves that in addition to increased national security, these "renewable" fuels will be a boost to agriculture and rural economic development; that they will result in improvements in environmental and public health; and that they will serve as an aid for international development.

The ugly truth of the matter is that not only are biofuels aka agrofuels not as renewable as we are led to believe (43), but "many of the biggest agrofuel boosters are familiar opponents to those now struggling for global justice, food sovereignty, and land reform. At the top of the list one finds such names as: ADM, Cargill, Bunge, ConAgra, Dreyfus, DuPont, Syngenta, Monsanto, Marubenji, Tate & Lyle, Wycrhauser, Tembec, British Petroleum, Misui, Royal Dutch Shell, Chevron, Mitsubishi, Petrobras, Total, Barclays, Morgan Stanley, Goldman Sachs, Societe Generale, and the Carlyle Group." (44)

In 2008, over 40 opposition groups involved in environmental, food and social justice issues publicly called for a moratorium on agrofuel production. (45) While this has not resulted in the anticipated moratorium it has at least raised public awareness of some of the hitherto undiscussed problems with agrofuels.

Unfortunately for these groups, not to mention planet earth, the U.S. military is making the mountain that much higher because for whatever misguided reason our military now assigns teams with the task of introducing agribusiness techniques to countries it is "liberating." (46) This project, like so many similar endeavors conducted by or on behalf of bank-beholden governments, tragically places our military squarely on the side of Big Ag and against the people and the planet.

Without question, growing fears over global warming has fueled the extraordinarily rapid concentration of the agrofuel industry. This concentration, comprised as it may be of seemingly unlikely climate change coalitions, has in no small part been driven by renewable fuel targets set by the U.S. and the EU together with other countries that fall under the OECD (Organization for Economic Co-operation and Development) umbrella.

Importantly, these are mandatory targets, reenforced by legislation, subsidies and various incentives – creating the legal matrix under which modern civil societies operate. As just one example, the U.S. Renewable Fuels Standard mandates growth in annual sales of agrofuels through 2022. All is driven by the CO2 bogeyman essentially created by economists as a tool by which to reduce negative externalities aka industrial pollution.

It is not small farmers, or we the people who come out on top in these situations. Even worse, current efforts to address climate change concerns through measures such as these not only ignore the urgent and very real biological needs of the soil but fail to take into account the impact of land use, and abuse, on climate. (47) Instead, our hubris and boundless faith in technology opens the door for various and sundry climate remediation projects, which in turn lobby for carbon taxes as means of financing opportunities and government subsidies for climate remediation projects.

Not-So-Strange Bedfellows

While agrofuels are an obvious fit for Big Ag and its support industries in biotech and petrochemicals, it does seem a bit odd, at least on first blush, that Big Oil would consider a move into agrofuels. Massive popular support for green energy has most certainly paved the way for oil companies to look for ways to incorporate renewable/clean energy projects into their business models. Fortuitously enough for the privately owned international oil companies in particular, "Big Oil's key assets—giant pipes, huge swaths of land with existing industrial-use permits, geo-engineering expertise, and extremely deep pockets—are 'tailor-made for developing and delivering biofuels.'" (48)

Although companies like BP and Shell have for years "had a scattershot approach, investing across the entire clean-energy spectrum," the sharper investment focus that has more recently emerged has proven to be a big win for agrofuels which also benefited from the vertically integrated cartels of Big Ag. No surprise then that some of Big Oil's biggest investments have been in the biotech arena, which is a key asset of Big Ag. (49) (50) This is an obvious – and for many a wholly unwelcome – boon for all things biotech, most notably genetically modified crops which are now creating superweeds and superbugs, together with a litany of devastating health problems associated with GMO foods and cotton. (51) (52)

These kinds of alliances between Big Ag, Big Oil and Biotech are actually restructuring agribusiness itself, with much of the money increasingly going into the construction of ever more fully integrated networks, through which production, shipping, processing and distribution are funneled. Moreover, "new, powerful players are converging into the [agrofuel] sector. Cosmetics corporations are selling biodiesel. Big oil is buying up plantations. Wall Street speculators are swinging deals with feudal sugar barons. All of this money circulating around the globe is reorganizing and intensifying transnational structures, linking the most brutal landowning class of the South with the most powerful corporations of the North." (53)

The virtually unsolvable environmental and social problems associated with the production of agrofuels has in no way interfered with its forward momentum. Indeed, growing opposition to agrofuels has only encouraged Big Oil and its allies to trot out "second generation" products, in the unfounded belief that industry-funded technology will *someday* provide the perfect solution.

Cellulose and algae are the current darlings of so-called second generation agrofuels. But, as in first generation agrofuels, they too come with serious, and related, downsides for peak soil concerns.

For example, cellulosic biomass (so-called agricultural waste) "requires large cropland areas, large quantities of water, and sunlight for green plant production [as well as] . . . . major fossil energy inputs." (54) Similarly and while algae's presumed benefit is that it uses less land mass, recent research has shown that algae production actually uses more water than other agrofuels. It also consumes more energy, due both to the process required to convert algae to fuel and its high fertilizer requirements (which contributes greenhouse gas in the form of nitrous oxide). (55)

Algae actually acts as an overly effective carbon sink, so much so that proper growth can only occur when CO2 supplies are adequate. One investment analyst wraps up the case against algae (and other agrofuels) nicely:

> "Algae can grow quickly, but only in the presence of sufficient nutrients. Just like any other organism, algae require carbon, nitrogen, phosphorus, and various other minerals. Algae can obtain carbon from atmospheric carbon dioxide, but the amounts present are insufficient to promote rapid growth. That requires

> something like smokestack effluents containing more than 10% CO2. . . .
>
> And when one is dealing with the other essential nutrients, which would likely be derived from solid wastes or agricultural residues, then one is competing with other forms of alternative fuel production methods that don't require photosynthetic life forms. . . .
>
> ….. Finally, the pervasive use of algal biofuel—and this is true of other biofuels as well—rests upon the assumption that the total biomass on the planet can be significantly increased or that spare capacity exists within global biosystems to support not only tremendously expanded cultivation of food crops but an approximately equal allocation of biomass to fuel applications. . . .
>
> Certain forms of algae [genetically modified?] may yet prove to be excellent high yield fuel crops. I'm betting against algae emerging as the perfect solution to our fuel problems though.
>
> At the same time I believe that *some people are going to make a lot of money in this new industry, not so much by selling fuel as by selling equity and by selling intellectual property."* (56) [italics mine]

As the above analyst implies, we can safely assume that five or ten years down the road, "when it becomes clear that algae fuel is devouring our water and energy resources and cellulose ethanol is mining our agricultural soils, we will be promised that the third generation will again solve all the problems of the previous generation." (57)

Money – that is to say, the promise of money – beckons and we blindly follow.

Decade after decade of trial-and-error experience should have taught us by now that our misplaced, unquestioning faith in technologies that do not heed the laws of nature will only serve to keep us on the path to certain destruction of the planet. In turn, this trajectory will continue to prevent us, as it long has done, from moving in the direction we need to go: toward a world of cheap, clean, abundant AND decentralized energy in which clean energy and Peak Soil need not be approached in terms of the current zero-sum game.

5

EXPLORING LAND-FRIENDLY ENERGY OPTIONS

BP and Royal Dutch Shell are two of a half dozen high profile oil giants which together make up a rarified group of international oil conglomerates, known as IOCs for short. These conglomerates characteristically organize into tightly managed systems of vertically integrated cartels which enables them to draw from an array of relatively diverse profit streams that now includes "renewables."

Their sheer size and economic importance allows them to exert incalculable influence both within individual governments and on the global geopolitical scene as well, factors which effectively deny the right of local people and communities to create and implement their own energy solutions. This situation almost guarantees their continued growth and influence.

Yet these IOCs are not without serious competition. Breathing down their backs are newly emerging state owned companies (or NOCs) which together control 52% of global oil production and 88% of global oil and gas reserves. The NOCs are located in countries around the world, including China, Vietnam, Russia, India, Mexico, Libya, Kenya and other African countries, Venezuela and other South American countries, and the Middle East. (1)

Some of these NOCs are moving toward privatization, voluntarily and otherwise. Russia for example is more or less voluntarily moving toward privatization of its NOC while Libya and Iraq's NOCs have been privatized as a result of war. But whether trending toward privatization or not, all of the NOCs have themselves been hard at work adopting management strategies that mimic those of energy giants such as BP and Shell so as to maintain economic viability, and supply much needed governmental revenue.

One such strategy is aimed at active participation in the international financial markets, or as some would say, the global casino. Climate change fears have provided the impetus through which BP, Shell and similar energy behemoths, together with NOCs have increasingly been placing substantial bets on the renewables/clean energy sector – much of which relates to projects designed for the carbon markets. (2) At the very same time, the free market is used as justification for allowing other problematic alternatives, such as tars sands oil to expand irrespective of known environmental costs.

The sanctity of the free market is in fact precisely why one of the U.S. State Department's latest environmental impact reports essentially holds that since tar sands oil is going to be mined, transported and burned anyway, building the hugely controversial, heavily contested Keystone XL pipeline is an economically preferable alternative to more expensive rail transport. In this respect, the State Department is right. Whether the pipeline is approved or not – and it more than likely will be at some point – the pipeline itself is merely an extension of the already established pipeline system. In the meantime rail transport serves to move tar sands oil, albeit more expensively and with just as much risk.

Heavy speculative investing is of course taking place in tar sands oil, but surprising as it may seem, heavy gambling by the oil giants is also taking place in the biofuels/agrofuels-related sector which provides a nearly perfect fit for Big Oil's already established infrastructure of pipelines, geo-engineering expertise and so forth. Unfortunately, and as we shall learn in Chapter 7, the zero-sum game dominating modern economic and financial thinking essentially guarantees that the wealth of the various participants in this "game" is continually shifted into fewer and fewer hands, in whose hands political power similarly accumulates. At the same time, investors and financiers alike ignore the fact that real wealth can only be created via land, and not through the stock markets, which (under the current monetary/legal system at least) see land and its products purely as commodities upon which financial wealth can be pyramided.

As a purportedly clean alternative, agrofuels have been seen as a panacea for the cap and trade markets, which are set to reignite with California's January 2013 entrance. Even worse and in appalling disregard of the still unfolding sub-prime debacle a key portion of the so-called carbon markets have been aimed at derivatives, "whose value is derived from the value of an underlying commodity – in this case, CO2 and other greenhouse gases." (3)

Thus, as former merchant banker-turned-novelist Linda Davies explains, these kinds of derivatives are essentially "bookie transactions once removed, taking a bet on a bet." In other words, their value is derived from fluctuations in the value of an asset, rather than from the asset itself, a fact which makes them tailor made for the carbon markets. Yet, such derivatives create an enormous, critically important disconnect between real wealth that ultimately only comes from the land, and financial wealth that not only all too often comes from irrationally exuberant speculation, but which represents mere claims on wealth, not wealth itself.

Push-back against carbon-based derivatives has come from a wide range of sources and include well known global warming activist James Hansen together with environmental groups such as Greenpeace and Friends of the Earth, who warn that carbon derivatives trading could not only increase carbon emissions but could also "trigger a financial collapse like the sub-prime loans crisis." (4) (5) Government officials and top experts have also raised concerns and objections (6) but unfortunately this casino may be in little danger of being shut down, given the sheer volume of money already invested and the political clout of those propelling it forward. Rather, rising fears over global warming and exploding debt have only increased the lust for simplistic solutions and quick profits, externalities be damned.

The Rules of the Game Puts Profits 1st, People & Peak Soil 2nd

As things stand now, and despite the growing chorus of objections from a wide spectrum of individuals and groups, it is likely that irrationally exuberant speculation in the renewable/clean energy sector will only become more entrenched as the tailor-made cap and trade derivatives markets slowly begin to make important legislative inroads. (7) This kind of gambling is zealously encouraged by financiers who are not likely to recognize or acknowledge the fact that this is the very kind of activity that creates a perpetually growing river of money and power flowing toward technologies and land use practices whose true costs are never actually accounted for, and which in the very near future spell the end of life as we know it on this planet.

Unfortunately, all is the direct result of our current monetary (or economic) system, in which our money is created as credit (or out of debt) by the banking system.(8) This monetary system, handed down to us from ancient Mesopotamia, is today following what should by now be an all too familiar pattern, whereby the most aggressive source of investment in renewable and less-than-renewable forms of "clean" energy is coming from the secretive and needlessly complex world of finance. This world includes mammoth banks such as Barclays and Société Générale, and giant equity funds such as Morgan Stanley, Goldman Sachs and the Carlyle Group, as well as politically powerful billionaires such as George Soros, Bill Gates and Vinid Khosla (of Google fame). (9)

Not surprisingly, the U.S. has positioned itself at the head of the Clean Energy/Climate Change pack. This is due both to its key position in world financial markets and its uber-strong lobbying system where, curiously enough, clients of the major climate lobbyists include the likes of the American Petroleum Institute, Chevron, Shell, Exxon Mobil, BP America, Sunoco, Duke Energy, Energy Future Holdings, Progress Energy, GE, Pacific Gas and Electric, Ford Motor Co, and more. (10)

Not to be left in the dust, clean coal joined the big climate lobbyists in 2008 under the banner of the American Coalition for Clean Coal Energy. ACCCE (pronounced ACE) is a "collection of 48 mining, rail, manufacturing and power-generating companies with an annual [promotional] budget of more that $45 million – almost three times larger than the coal industry's old lobbying and public relations groups combined." Dwarfed by this PR maelstrom was the counter-charge formulated by the ever-vigilant Greenpeace, objecting to "'the untold billions of dollars in handouts for the false promise of carbon capture and sequestration." (11)

This confluence of competing interests is of course dominated by Big Money (big banks and big business), which however well intended, is able to steer public energy options towards those that are best for its own bottom line and not toward the needs and desires of local communities. As a result, oxygen-producing rain forests in the Global South are being razed and indigenous populations are being pushed aside to make way for biofuel plantations, hydropower dams, and oil fields. Meanwhile the landscape across the U.S. and Global North is being decimated by tar sands oil and shale gas extraction, windmill farms, ethanol plants and weather modification experiments. All is reinforced by specific language found in virtually all of the various, mis-named free trade agreements.

Lost in all the noise and human angst is any discussion of politically incorrect but potentially planet-saving alternatives.

Mentioning the Unmentionable

Oddly enough, there is one politically incorrect and therefore unmentionable biofuel that just might work as a viable alternative to current agrofuel choices, at least insofar as Peak Soil issues are concerned. That is of course Cannabis sativa L. or hemp, which, up until about 100 years ago, had been for some 6000 years prior widely used as a key food and fiber crop. At the time of the American Revolution, hemp was used to pay taxes, and hemp was used to make the sails of ships. The founding documents were also printed on paper made from hemp. By the mid-1850's Levi Strauss was making miner's jeans from hemp canvas. In the 1890's farmers were using home grown hemp oil as fuel for their new tractors and during WWII farmers were encouraged by our government to "grow hemp for victory."

According to the research of renowned herbalist James A. Duke, taken from his 1983 unpublished *Handbook of Energy Crops* (12), non-genetically modified Cannabis sativa L [Cannabis hemp] is "a multiple-use plant, furnishing fiber, oil, medicine, and narcotics. Fibers are best produced from male plants. In the temperate zone, oil is produced from females which have been left to stand after the fiber-producing males have been harvested."

Duke also discusses cultivation of Cannabis Hemp: "As *Cannabis sativa* has been cultivated for over 4,500 years for different purposes, many varieties and cultivars have been selected for specific purposes, as fiber, oil or narcotics. . . However, all so called varieties freely interbreed and produce various combinations of the characters." Additionally, Duke says hemp plants are propagated easily by seeds and are very adaptable to soil and climate conditions, requiring little cultivation except for weeding during the early stages of growth.

Running contrary to popular belief, Duke says that hemp "tends to exhaust the soil of nutrients." While acknowledging that some nutrients are returned back to the soil after these plants are harvested, Duke makes some organic suggestions for soil dressing, especially needed for fiber-producing plants.

This being said, and in direct contrast to the far less vigorous "industrial hemp" which has recently been gaining a modicum of official support, the deep root system of Cannabis Hemp does actually aerate and help replenish the soil to a much greater degree than nearly all other plants, including switch grass. (13)

Duke also points out that "In many areas several varieties of hemp are grown so as to spread out the harvest, one maturing in late July and used later for seed crop in September, a second crop maturing in mid August, and a third maturing near the end of August." In other words, Cannabis Hemp requires very low inputs – that is to say little or no petrochemical fertilizers or pesticides, no genetically modified seeds and little or no irrigation when grown in appropriate soils and climates – in order to reap four sizable harvests per year on the same land.

This kind of cultivation is by far and away best accomplished by small farmers and not Big Ag, whose economic model requires extracting whatever one can out of the soil as well as monopoly control of patented seeds, regardless of the social and environmental costs. Pulling weeds, seed saving, soil dressing with manure from healthy, clean animals, and selecting appropriate soils and climate are simply too expensive, restrictive and labor intensive for Big Ag operations which must meet profit expectations of stockholders and lenders in addition to finding ways to manage debt obligations.

In addition to producing cellulosic fiber for conversion to biofuel, Cannabis Hemp, as Duke details, is a very versatile crop that can be used to make an amazing array of useful products. These include high quality textiles, medicine, folk remedies, food, and biodegradable plastics. All of these products can and should be produced using environmentally friendly, low impact techniques. This concern for the environment and the soil should of course extend all the way from the production to the disposal process.

Indeed we find that plastics made from hemp are – unlike their petrochemical cousins – fully biodegradable. And textiles made from hemp are universally far gentler on the environment than their synthetic cousins, or for that matter, resource-hungry, soil-damaging cotton. For example, Dresden Cotton made from Cannabis Hemp can serve as viable, marketable substitute for soil-destroying cotton. Similarly, plastics made from Cannabis Hemp can put an end to "mermaid tears" now choking our oceans slowly to death with non-biodegradable oil-based plastics. (14) (15) (16)

In addition, depending how it is grown and harvested, Cannabis Hemp can produce both oil for food and cellulosic fiber for conversion to biofuel. Moreover and using the ancient pyrolysis technique it can produce "the variety of liquid fuels and lubricants type-similar to gasoline and subterranean oil, but without pollution." Some even claim that if the U.S. National "Soil Bank's" registered fallow land (about 89 million acres) were to be planted in non-gmo Cannabis hemp, we could produce "enough Cannabis-Methanol, oil, lubricants and BTU-gas for all North American land, sea and air transportation requirements." (17)

While this may be a bit of an overstatement, particularly if hemp's food and fiber uses come into serious competition with its biofuel potential, and if proper attention is paid to maintaining soil quality, it does nevertheless appear that Cannabis Hemp holds real promise in helping to rebalance the energy/Peak Soil equation.

In an ideal world, our duly elected leaders and allied experts, along with the citizenry as a whole, might at least explore the amazing potential of Cannabis Hemp to fill at least a sizable portion of our energy, medical, clothing and packaging needs while at the same time helping to rehabilitate the soil through its deep-root system. So far that isn't happening in any serious way, other than to periodically trot out the overly narrow solutions we know as industrial hemp, medicinal marijuana and legalized "pot."

So you ask, what's left? The answer unfortunately is even more politically incorrect than non-industrial Cannabis Hemp, and its pursuance will entirely depend on how serious you and I are about solving our energy needs in such a way as to preserve our precious soils, our most basic of human rights – and the planet itself.

Thinking the Unthinkable: Abundant, Clean, Cheap, Decentralized Energy

Too few of us can imagine a decentralized form of energy that can be produced and delivered locally, and even within individual households. Those that can imagine such an energy source will limit their thinking to small scale wind, solar, or biofuels, which by definition are not cheap, clean or abundant and which by their nature can only serve isolated aspects of our energy needs.

We are, nevertheless, closer than we might ever have previously dared hope to such incredible forms of energy. This energy falls under three general categories, including one known as cold fusion, but generically it is called new energy, or somewhat less accurately, "free" energy.

The history of cold fusion is more or less illustrative of what has occurred elsewhere in new energy research over the past several generations. The story perhaps begins with the term cold fusion, which was originally used in the scientific literature to describe an altogether different process from what the term has come to mean today. The term is, as a result, somewhat misleading.

In the more general sense, researchers define the cold fusion process as "the generation of anomalous excess heat at low temperatures." (18) However, other scientists working in the field believe that because cold fusion is such a misconstrued term, other more accurate terms should be applied. They suggest that "the names 'Chemically Assisted Nuclear Reactions' (CANR) and 'Low Energy Nuclear Reactions' (LENR) more correctly describe the phenomenon." (19) (20)

Despite all the negative press, truly clean, cheap, widely available, decentralized energy derived from the mis-named cold fusion process is quite likely to be much closer to becoming a present day reality than is commonly assumed. For example, a 2009 report titled "Shaping the Energy Technology Transition" was published by the Policy Research Project on Energy Technology Transition under the auspices of the Lyndon Baines Johnson School of Public Affairs. Among its findings were the following:

> "Two case studies in this report—biofuels and cold fusion—demonstrate that current energy policymaking has not always led to an optimal solution for maximum public welfare benefit. The biofuels case is viewed by many as an example of energy policy that did not necessarily take into account all relevant economic and environmental factors, such as water supply, soil erosion, and increased use of fertilizers and pesticides.
>
> Cold fusion represents a case in which negative public policies, *despite a continuing possibility that the phenomenon may eventually be found to be 'real,'* may not necessarily serve the long-term public interest." (21) [italics mine]

The term cold fusion, first used in the scientific literature in the late 1980's by Physicist Steven E. Jones and his team at Brigham Young

University to describe their own experiments, unfortunately became instantaneously and erroneously associated with work of Drs. Stanley Pons and Martin Fleischmann which was rush-released in 1989 amidst a storm of controversy. According to the online *New Energy Times*:

> "The process discovered by Jones' team is markedly different from the process discovered by Fleischmann and Pons. . . .[W]hat Martin Fleischmann and Stanley Pons discovered . . . was that, in their cold fusion cell, Q, the amount of heat energy coming out of the cell was up to 1000 times greater than it should have been based on any known chemical reaction.
>
> An excessive amount of heat was coming from the experiment. It did not, in any way, match the amount of electrical energy going in plus other accounted-for energy losses. And this, in a nutshell, was their fundamental historic discovery: something within the cell was releasing a new, "hitherto unknown" (Fleischmann-Pons) source of potential energy. In LENR research, this is the most important aspect of the phenomenon and is known by the term "excess heat."
>
> . . . Fleischmann and Pons introduced an entirely new field of science. It didn't belong to physics; it didn't belong to chemistry. It was somewhere between them. A turf battle started the day it was announced." (22)

The Fleishmann/Pons controversy rages on today with erstwhile supporters scrambling to piece together funds and scientists willing to put their careers on the line in order to continue research in the field.

Yet, as the aforementioned report on "Shaping the Energy Technology Transition" stated, negative public policy on cold fusion presents a serious impediment to such research despite the continuing possibility that the phenomenon may be found to be real. However, said public policy has been but one of several impediments which include Big Energy, the military and academia. Taken together, these form a formidable barrier to further exploration and development of cold fusion and other forms of new energy. Counterintuitive though it may seem, the scientific community itself has, in the main, created its own numerous but effective barriers, as the late Eugene Mallove inadvertently discovered once he had become a cold fusion proponent, and later a New Energy proponent.

Mallove's impressive list of credentials include two degrees in engineering from MIT and a doctorate in Public Health from Harvard. Articles written by and about him appeared in *MIT Technology Review*, *Popular Science*, *Wired*, *The New York Times* and elsewhere. In addition to being a chief science writer for MIT, he appeared as a guest on numerous national radio programs and was, during the mid-1980s, a regular science and technology broadcaster for the Voice of America.

As a result of his investigative reporting, Mallove wrote three critically acclaimed science books for the general public. He resigned from his post as chief science writer at MIT in order to write his third book which was published in 1991. It was called *Fire from Ice: Searching for the Truth Behind the Cold Fusion Furor.*

Not only was this book nominated for the Pulitzer Prize, but it received the following endorsement by Julian Schwinger, Nobel Laureate in Physics (1965): "Eugene Mallove has produced a sorely needed, accessible overview of the cold fusion muddle. By sweeping away stubbornly held preconceptions, he bares the truth implicit in a provocative variety of experiments."

Mallove himself wrote that he was "most proud of this latter book, because it began a jarring quest that led to finding out not only dramatic new truths about new accessible forms of energy in nature, but a most astonishing truth about modern 'official' science: Official science is not really intent on truly expanding scientific knowledge, in particular when some very, very fundamental scientific dogmas and theories are put at risk."

Mallove's investigation into the attempts of MIT nuclear physicists to replicate the original Pons-Fleischmann work led him to the unhappy discovery that these scientists in fact fraudulently reported their inability to replicate the Pons-Fleischmann experiments. In fact, Mallove found that their *unadjusted* findings *actually reinforced* the veracity of the Pons-Fleischmann discovery. Mallove quite logically surmised that part of the reason for this fraud was that these physicists had a vested interest in hot fusion research – research that was then, by itself, consuming tens of billions of federal dollars.

In 1995 Mallove founded the small but highly respected *Infinite Energy Magazine* (23), which he edited until his untimely death in 2004. The day before he died, Mallove wrote "an open letter to all people of the world who have open-minded curiosity, good will, good judgment, and imagination."(24) Mallove's letter was a plea for support in order that current findings about new energy be brought to fruition. Surprisingly, money, though needed, was not the major obstacle because despite the fact that *"many, many peer-reviewed technical publications have indeed courageously published pioneering technical papers about new energy,* [the major stumbling blocks have been] the prominent mainstream publications that set the boundaries of the public scientific discourse. [Thus] journals such as "Science and Nature" reject *without review* any and all papers that challenge the foundational paradigms of physics, chemistry, and biology." [italics mine]

In this same letter Mallove asserted that those experts who claim that several hundred billion dollars are needed to achieve viable, environmentally safe energy alternatives are simply wrong. Rather, he said, "I have news for these experts: The solutions to our energy problems are very close at hand, and they do require initial research and funding, but not the billions of dollars that such Establishment "experts" are accustomed to from government largesse. *Rather, all that is needed perhaps are only several tens of millions of dollars to create robust prototype electric power generators based on new energy physics discoveries that have already been made."* [italics mine]

While Mallove's *Fire From Ice* led him to become perhaps one of the world's foremost proponents of cold fusion energy (and later other forms of new energy as well), it was a self-taught Austrian by the name of Viktor Schauberger who is considered to be the father of cold fusion. (25) For those familiar with his work, he is one of the unsung geniuses of our age.

Unfortunately, his unorthodox, if highly ingenious, methods won him little besides a marked resistance from the received knowledge of academia. This is slowly changing, and today we find that a number of his ideas have been put to practical use. As just one example, the Austrian water engineer Otmar Grober won the Stryrian Environmental Award in 2001 for his work in re-regulating a number of rivers based on the principles of Schauberger. (26)

Schauberger is best known as an inventor, naturalist and highly respected ecological pioneer who developed the Golden Plough and other methods for revitalizing the soil. (27) (28) According to Alick Bartholomew, author of *Hidden Nature: The Startling Insights of Viktor Schauberger*, Schauberger excelled as a teacher of the science of Nature, describing and illustrating as few have done how Nature works. "His aim was to liberate people from dependence on inefficient and polluting centralized energy resources and generation of power." says Bartholomew. Bartholomew adds that:

> "From his astute observations he became a self-trained engineer, eventually learning, through the implosive, or centripetally moving, processes which Nature uses, how to release energy 127 times more powerful than conventional power generation. By 1937 he had developed an implosion motor that produced a thrust pf 1,290m/sec, or about four times the speed of sound. In 1941 Air Marshall Udet asked him to help solve the growing energy crisis in Germany; however the research came to an end when Udet died and the plant used was destroyed by Allied bombing. . .
>
> The critical tests came just before the end of the European war. A flying disc [developed under threat of death for the Third Reich] was launched in Prague on February 19, 1945, which rose to an altitude of 15,000 meters in three minutes and attained a forward speed of 2,200kph. An improved version was to be launched on May 6, the day the American forces arrived at the Leonstein factory in Upper Austria. Facing the collapse of the German armies, Field Marshall Keitel ordered all the prototypes destroyed." (29)

Schauberger's early observations and meticulous study of water led him to his exploration into energy alternatives and nature-respecting inventions. Viktor used his observations about the properties of water and the way in which it interacted with the surrounding environment to "initiate designs for generating power and motion that worked in harmony with nature. . . For Viktor nature is the foremost teacher so that the task of technology is not to correct nature but to imitate it, 'kapieren und kopeiren' was the principle that guided him throughout his life, 'first understand nature and then copy it." (30)

Between 1922 and 1928 Schauberger developed a method of building log-flumes, for which numerous patents exist, that would transport heavier-than-water timber over long distances by maintaining the temperature of the water in the flumes, and by inducing a spiraling motion in the logs being transported.

Schauberger's original proposal was initially met with derision by experts and academics because his methods, which heeded the laws of nature, went completely against accepted hydraulic principles. However, once completed, his first project was such a resounding success that experts from all over Europe came to study the flume's unique construction, and his fame spread.

Academics and experts again became incensed when Schauberger was appointed State Consultant for Timber Flotation. Not only was he paid twice as much as they, but he was allowed to give directives on technical questions for which he had no credentials.

A crisis ensued, but because his flumes substantially reduced the costs of bringing trees out of inaccessible forests, with no damage done to the timber, the expanding building industry welcomed Schauberger's invention with open arms. Schauberger resigned his State position, and took a job with one of Austria's largest building contractors, and built successful flumes all over Europe.

His inventiveness didn't stop here, though if it had he still would be known as the man who mastered the art of the previously impossible: economically transporting timber by water, without doing damage to the timber.

As it happened, the years following World War I also brought about significant changes in land use. These changes reflected the urgent need for revenues, and led to widespread deforestation. As had happened in Mesopotamia long ago, deforestation led to horrific levels of soil and river degradation. Schauberger saw that conventional soil and water management policies actually made problems worse and this led him to investigate and introduce alternatives which again were designed to work with nature instead of against it.

Petitioning the Bonn government to re-establish the natural flow of the Danube and Rhine represented one of his campaigns. Another was to develop a method to generate electricity that was less destructive than hydroelectric dams. He achieved a measure of success in the early 1920's when he developed a method through which 90% more electricity could be generated from a given flow-volume of water than could be achieved by conventional methods. According to Alick Bartholomew in his book *Hidden Nature* Schauberger installed this device, using water from a nearby stream, to light his forest warden's house, which was too remote to be connected to any other source of supply. In 1929, he patented a winding-impeller made out of copper, for electricity production out of a river, under the name Jet-Turbine. (31)

During the next several years Viktor continued his experiments for producing electrical energy directly from water, and he also investigated methods by which degraded water could be converted to pure, life-giving water. In addition, he "investigated a new motor fuel that could be used in ordinary combustion engines, but without the dangerous waste products. He discovered that water in a constructive hyperbolic motion has the ability to bring about the syntheses of hydro-carbons suitable for fuel. In his experiments 'water sprayed into a cylinder and a quantity of natural oxygen is added, a light heat pressure created by a descending piston is sufficient to transform the highly potent water into gas.'" (32)

Also during the 1930's Viktor undertook a study of comparative agricultural methods, sanctioned by King Boris, and in 1933 he published his only book titled *Our Senseless Toil – The Cause of the World Crisis,* with the subtitle *Growth through Transformation, Not Destruction, of the Atom.* In 1938, and in partnership with his son Walter, he constructed a replica of Lord Kelvin's Falling Water

Experiment of capillary research, generating 20,000 volts of electricity. (33)

In 1934, Schauberger met with Hitler to discuss the fundamental principals of forestry, agriculture and water engineering. Hitler was impressed enough by Schauberger's radically different ideas for utilizing water power in new and dynamic ways that he wanted every possible assistance given to Schauberger's research into fuel-less energy production.

Apparently however, Schauberger declined to work for the Third Reich. This did not bode well for him after Austria was annexed to Germany in 1938. When war broke out in 1939, the SS would soon come searching for him and his ideas based on his 1935 patents for an air turbine and procedures for lifting liquids and gases, and the "warm-cold" machine, called the Klimator, built for Siemens in 1937 (which was, however, destroyed in an unauthorized test). (34)

By 1940 he had conceived, patented and built various versions of the repulsine motor, using spin and suction-based implosion-technology-based technologies for the purpose of energy-creation which would be used for such things as heating or cooling houses. The first Repulsine, more commonly known as the flying saucer, was built first in Berlin and then in Vienna, where the prototype broke from its mooring and smashed through the factory's ceiling.

Just prior to this, Professor Ernst Heinkel, who had himself developed an innovative aircraft that bore his name, stole Schauberger's confidential patent application and unsuccessfully attempted to incorporate it into his new jet aircraft, the HE 280. After determining that his own conventional aircraft frame was completely inappropriate for Schauberger's suction engine, Heinkel developed and built his Schriever 'Flying Top' in his Rostock factory in 1940. (35) By 1943 refinement work on the Schriever saucer was moved to a secret location in Czechoslovakia, to which Schauberger was periodically escorted.

This same year, Himmler gave the German SS responsibility for the development of secret weapons which drew upon Schauberger's innovations. Production facilities were set up in giant cave complexes in Poland and Czechoslovakia. Although claims have been made that Schauberger himself built flying saucers *this has never been verified.* Research does indicate that "Viktor's intention was to develop a totally new method of propulsion. In the 1940s he experimented with rotating waviform shape discs. *These devices were designed for practical purposes such as heating or cooling houses (the "Klimator") or for propulsion.* In planes and submarines, his engines were designed to create a 'biological vacuum' in front, which should suck the vehicle forward. . . But as far as we know, Viktor only experimented with small discs, less than 1 Meter in diameter, as mentioned." (36) [italics mine]

Nevertheless, the level of secrecy was so high at the SS research complexes "that at one crucial location at the end of the war, the SS lined up 62 of the scientists and laboratory technicians, and shot them, to save the secrets of that complex 'free energy' atomic installation from falling into the hands of the approaching Soviet troops." (37) At the same time, Field Marshall Keitel ordered all prototypes and working models of Schauberger's Repulsine destroyed.

Beginning in the 1930's Viktor had also begun work on a prototype power plant that would extract energy directly from air and water. The first experiments on this prototype took place during 1931-32, and were undertaken with a Viennese engineer by the name of Dr. Winter. When these experiments proved inconclusive, Schauberger went back to his drawing board. By 1955 he had developed what became known as the Implosion – or trout turbine – machine which the anti-nuclearists promoted as an alternative source of power.

Though Viktor was still perfecting his design for the "trout turbine" machine, he received so much publicity that he was approached by an American consortium led by Karl Gerchsheimer and Robert Donner who extended an invitation to come to the US to continue development of his inventions. After spending several months writing articles and drawing sketches, Viktor became disillusioned and suspicious about the true motives of his US partners and decided to return to Austria. In order to do so, he was forced to sign an agreement which forbade him to share any of his work or writings with anyone other than Mr. Robert Donner. He also had to sell all his work and patents to his American partners before he could buy passage home. Within five days after returning to his beloved Austria, Viktor Schauberger was dead.

Years later a Russian scientist by the name of Evgeny Podkletnov drew upon original Schauberger papers, which had been acquired by his father (who was himself a leading authority in hydro-engineering), to build an anti-gravity device. In July of 2002 the venerable *Jane's Defense Weekly* announced that they had seen secret research papers from Boeing. These secret research papers revealed that Podkletnov's anti-gravity device had indeed been introduced into the Boeing Aerospace program. Like BAE Systems and Lockheed, Boeing had even contacted Podkeltnov directly for more details and help with their research. All three companies were reported to already "have some activity in this area of research" which went under the more politically correct term "propellentless propulsion" instead of anti-gravity.

Acknowledging that this research "could rewrite the economics of conventional aviation technologies" Boeing stated that "[i]f gravity modification is real, it will alter the entire aerospace business." Among the more chilling revelations in this report was the conclusion that one key goal of on-going research "has been the need to examine possible uses of anti-gravity technologies. *Conceivably, uses could range from transport both terrestrially and in outer space through fuel-free electricity to powerful directed energy weapons.*" [italics mine]

Research was partially stymied when Boeing "fell foul of Russian technology transfer controls." But perhaps even more importantly and like post-WWII Schauberger before him, Podkletnov was only willing "to provide assistance if the research was conducted out in the 'white world' of open development." Nevertheless, "the briefing document [obtained by *Jane's Defense Weekly*] said additional classified activities in gravity modification may exist." (38)

While the technologies discussed in the *Jane's Defense Weekly* article above might seem unbelievable to even sophisticated readers, there was a time when the first attempts at flight by the Wright brothers might have looked more like flights of fancy than anything grounded in any kind of reality. Conversely, Nicola Tesla's famous experiments with electricity often seem more unbelievable to modern audiences than they were in Tesla's time due to the attention he garnered – in no small part as a result of his personal connections with such well known personages as Mark Twain and George Westinghouse. Yet since the time of the Wright Brothers and Nicola Tesla we have witnessed the development of the pilotless airliner, the stealth bomber, unmanned arial vehicles with bombing capabilities, the "death ray" or laser weaponry and countless other technological marvels we have come to accept as inevitable.

Making headlines in recent months are directed energy weapons or DEWs which were mentioned in the *Jane's Defense Weekly* article above. According to the Federation of American Scientists, directed energy weapons or DEWs include laser, as well as high power radio frequency (HPRF) which is frequently called high power microwave (HPM) or RF-Directed Energy – and particle beam technologies. (39)

In April of this year, a report was aired on ABC News discussing the Navy's new Laser Weapon System or LAWS which it said is no longer "the fantasy weapon of the future" since the Navy is "set to field a powerful laser that can protect its ships by blasting targets with high-intensity light beams." Strong enough to burn a hole through steel, "each pulse of energy from the laser costs under a dollar and it can be used against weapons systems that are significantly more expensive." The "secret sauce" used to develop and perfect the LAWS clearly demonstrates that the military and its business partners have a decided interest in seeing to it that such secret sauces remain just that – secret. (40)

In his 2004 "Open Letter," the late Eugene Mallove wrote that the research discussed in his *Infinite Energy* magazine "suggests that there are at least three major categories of radically new sources of energy that civilization is on the verge of being able to tap and reduce to practical technologies. These are the completely new forms of energy . . . that are currently not recognized as feasible by the 'scientific establishment,' but for which overwhelming and compelling evidence exists."

Cold fusion was the first of the three major categories that Mallove discusses in his Open Letter, as follows in part:

> "Category 1. New hydrogen physics (a.k.a. "cold fusion"), more generally Low-Energy Nuclear Reactions or LENR, "hydrino" physics, and other water-based energy sources. Copious technical and other information about this research may be found on these two diverse websites: www.lenr-canr.org and www.blacklightpower.com as well as our own site, www.infinite-energy.com. The upshot of this energy-from-water field is that within ordinary water there is a heretofore unimaginably large energy reservoir that may be as great as 300 gallons of gasoline energy

> equivalent within each gallon of plain water! This energy would be non-polluting, would have no hazardous radiation, and would, in effect, have a zero fuel cost. Only one cubic kilometer of ocean water would provide energy equivalent to all the known oil reserves on Earth. . ." (41)

The reader may recall that during the 1930's Viktor Schauberger was investigating a method by which water could be turned into hydrogen gas utilizing a "constructive hyperbolic motion" for use in ordinary internal combustion engines. There were two advantages to a device such as Schauberger's. First, the hydrogen gas could be created onboard and on demand, using any type of water, eliminating the problems shared with conventional gasoline of how to store and distribute highly inflammable hydrogen, factors which require expensive infrastructure including extensive pipeline networks and gas stations. The second advantage is the fact that when the hydrogen gas "burned" the only exhaust was water vapor.

It turns out that Viktor wasn't alone in his quest to create a water powered vehicle. On July 2, 1932 Henry "Dad" Garrett, the most famous Afro-American inventor of that time, together with his son Charles, filed a patent for what they called an electrolytic carburetor by which water was split into hydrogen and oxygen atoms, and mixed with air in an internal combustion engine. Son Charles became the first person in the United States reported to run an internal combustion engine with water, as reported by the Dallas Morning News, September 8, 1935. Although once accepted only as rumor, one determined investigator managed, with some difficulty, to obtain the patent the Garretts had for their invention, which said investigator prefers to refer to as a hydrolytic carburetor to better describe *what* is combusting while also commenting about his "total amazement at the simplicity of the technology." (42)

However, neither Schauberger nor the Garretts were the first to have come up with the idea of extracting hydrogen and oxygen from water which then could be used as a fuel. In 1806, a Swiss engineer and politician by the name of François Isaac de Rivaz invented an internal combustion engine that used a mixture of hydrogen and oxygen, obtained from water, as fuel. This engine, equipped with an electric ignition, was described in a French patent in 1807. The next year Rivaz designed a rudimentary "car" to go with his engine, but this was a commercial failure. (43)

Twenty-five years after Rivaz obtained his patent, the first electric cars were invented. Some fifty years after that, in 1885, the modern gas engine was invented. According to the World Intellectual Property Organization (WIPO), electric cars were more popular than gasoline powered cars at the turn of the twentieth century "for much the same reasons that consumers are taking a second look at electric cars today: they did not emit noxious fumes, were quiet, smoother and easier to drive." Gasoline powered cars, says WIPO, took over the market due to cheap, abundant gas, good roads and the low, low price of Henry Ford's Model T with which no other car could compete.

It is perhaps of some interest to note here, given our earlier discussion, that, as was reported in a 1941 issue of *Popular Mechanics*, the first Model T was actually built to run on hemp oil, although it also ran on gasoline, kerosene, moonshine or just about anything else you could put a match to. The car itself was made entirely of hemp whose resin-stiffened panels had an impact strength ten time that of steel. (44)

Once the infrastructure required for gasoline distribution was built, most Model T's were run on gasoline, then cheap and abundant. Some researchers suggest that readily available, cheap gas was a major factor contributing to the commercial failure of Garrett's invention – while others (including those who claim to have perfected the device) say that the Garrett model resulted in no net energy gain, requiring one unit of energy input for every equivalent unit of energy output.

Be that as it may, and as the WIPO additionally points out, neither electric cars nor the concept of hydrogen powered cars have ever completely disappeared – despite cheap gas prices. They "simply faded out of the mass consciousness until the 1970s gas crisis and environmental concerns brought them back to the fore." Today of course both hydrogen and electric powered vehicles are once again enjoying renewed public interest.

Two of the most highly touted advantages of both electric cars and hydrogen powered cars are that, firstly, they have zero-emissions and secondly, because they have few moving parts, they are quiet and vibration-free. Also acknowledged is the fact that "today's internal combustion engines can be readily converted to run on a variety of fuels including hydrogen. Hydrogen is one of the most plentiful elements in the universe. It can be extracted from natural gas, coal, crude oil, etc. but water is the only pollution-free source of hydrogen. The hydrogen and oxygen atoms on water can be easily and cleanly split apart by electrolysis. . . The resulting hydrogen can be compressed for storage and use in fuel cells." (45)

The question that remains however is that if "the hydrogen and oxygen atoms on water can be easily and cleanly split apart by electrolysis" then why should we focus solely on building vehicles that rely on compressed hydrogen, which require expensive transportation mechanisms, storage facilities and innumerable hydrogen gas stations? But besides this question, there are additional issues regarding conventional hydrogen fuel cell cars, which Eugene Mallove points out in his 2004 Open Letter:

> "Conventional hydrogen fuel cells, which are widely discussed by the news media today, rely on the conventionally understood energy from hydrogen

> when it combines with oxygen to form water. This is thousands to millions of times less powerful per gram of hydrogen than *already demonstrated* new energy sources! Furthermore, the hydrogen for conventional fuel cells must come from some other energy source that must be used to break down abundant water to get hydrogen fuel (if we reject the other hydrogen source: hydrocarbon fuel). *But in all conventional hydrogen fuel processes using water as the starting material, this requires more energy than one gets back when the hydrogen is consumed.* So ordinary 'hydrogen power' is a misnomer at best – it is no solution at all to the world's real energy needs. *Hydrogen, conventionally employed, is an energy storage medium – period."* [italics mine]

Industry is nevertheless moving ahead with "conventionally employed hydrogen." As reported by CNN Money this past August of 2013 we just may be "on the verge of a hydrogen moment" because "virtually every major carmaker is preparing to launch hydrogen-powered models in the next few years." One reason is that the auto industry may simply be "scrambling to meet California's strict air regulations" particularly since electric car sales are not living up to expectations. The biggest challenge facing fuel cell cars today, says the report, is the same as it's always been – a lack of infrastructure. This problem however may have been presented with a window of opportunity via today's lackluster job market, so long as the government provides adequate support. As the report explains,

> "By comparison, there are some 250 million gasoline-powered vehicles on the road in America, serviced by 170,000 filling stations. Each new hydrogen station costs as much as $3 million to build. A back-of-the-envelope calculation suggests it would cost $512

> billion – roughly the GDP of Sweden – to replace all the gas stations. . .
>
> The danger is that America may fall short on yet another green technology (like the rapid demise of U.S. solar manufacturers). Germany has committed to building 50 new hydrogen filling stations, and Japan to 100. It's largely up to Washington to make sure that in the U.S. hydrogen doesn't remain stuck as the fuel of the future." (46)

Meanwhile, new energy research is also moving forward. For example, a company by the name of Blacklight, mentioned in Mallove's Open Letter, "owns and occupies a 53,000 square foot modern research and development facility equipped with state-of-the-art laboratory equipment. The Company's technical core competencies are theoretical physics, molecular modeling, electrochemical system research, development and testing, characterization of chemical processes, thermal power measurement, and chemical, material, and plasma characterization and spectroscopy. Currently, the Company has twenty employees and fourteen consultants. The majority of employees are scientists and engineers, including six Ph.D.s." (47)

The following excerpts from Blacklight's "Business Summary" page provides an overview of the world's growing energy needs and how alternatives will fall short despite massive growth and investment – and it includes a few of its impressive and exciting developments to date:

> "According to the Energy Information Association ('EIA'), during the last 2 decades, global demand for energy has nearly doubled. This rapid growth has resulted in imbalances of supply and demand of electricity and natural resources worldwide. Global energy demand has grown 24.1% from 2000 to 2008,

and the EIA projects growth of 51% by 2035 from 2010 demand levels. Energy provided by existing renewable sources, primarily wind and solar which are relatively poor sources of baseload power, is projected to grow rapidly during this time period, but is expected to satisfy only 14% of global demand by the year 2035 (Figure 1). In fact, despite global mandates on the use of renewable forms of energy and increased energy conservation, consumption of fossil fuels is projected to increase 38% over the same time period. . .

In its facility in Cranbury, NJ, BlackLight Power, Inc. has a proprietary breakthrough power-producing system *continuously operating* that converts ubiquitous H2O vapor into electricity as the only source of fuel as confirmed by six separate independent individual or groups of leading scientists from academia and industry with PhDs from prestigious universities including the Massachusetts Institute of Technology and the California Institute of Technology. Specifically, BlackLight Power, Inc. *has developed a commercially competitive, nonpolluting fundamentally new primary source of energy that forms a prior undiscovered, more stable form of hydrogen called "Hydrino" predicted by its founder Dr. Randell Mills that releases two hundred times more energy than burning hydrogen. . .*

. . . *To date, the Company has licensed the rights to produce approximately 8,250 MW of new electrical power to seven companies, including five electrical utilities and two independent power producers.* Collectively, these utility companies own, purchase, or manage electric power production of approximately 7,600 MW and service nearly one million customers. The avoided fuel

> costs from these agreements could be in excess of $2 billion per year." (48) [italics mine]

As Mallove suggested, the years of research it took for Blacklight to develop their "Hydrino" technology took just $60 million – not the tens of billions so often thought necessary for more conventional avenues of energy research. But "Hydrino" technology is not the only "new energy" device to have made headlines recently. In 2008, the world was treated to headlines which read "Ocean Currents can Power the World, Say Scientists." The power source was a revolutionary new device conceived by scientists at the University of Michigan (which is one of the top ranked engineering schools in the nation.) (49)

The project was headed by Michael Bernitsas, a professor in the University of Michigan Department of Naval Architecture and Marine Engineering. The description provided by Bernitsas was strongly reminiscent of Viktor Schauberger's trout turbine technology:

> "VIVACE stands for Vortex Induced Vibrations for Aquatic Clean Energy. It doesn't depend on waves, tides, turbines or dams. It's a unique hydrokinetic energy system that relies on 'vortex induced vibrations'. . . VIVACE copies aspects of fish technology. . . Fish curve their bodies to glide between the vortices shed by the bodies of the fish in front of them. Their muscle power alone could not propel them through the water at the speed they go, so they ride in each other's wake. . . The very presence of the [fish-like] cylinder in the current causes alternating vortices to form above and below the cylinder. The vortices push and pull the passive cylinder up and down on its springs, creating mechanical energy. Then,

> the machine converts the mechanical energy into electricity." (50)

The University of Michigan obtained a patent for the VIVACE and a company by the name of Vortex Hydro Energy has obtained an exclusive license to commercialize it. As the Vortex Hydro Energy webpage states, the VIVACE converter is unlike water turbines in that it does not use propellers. Being a transformational technology it taps into a vast new source of clean and renewable energy, that of water currents as slow as 2 to 4 knots previously off limits to conventional turbine technology that target rivers with water currents greater than 4 knots. The vast majority of river/ocean currents in the United States are slower than 3 knots. Supporters include the National Science Foundation, the U.S. Navy and the U.S. Department of Energy among other. (51)

If these new energy forms can be fully realized it means that, as the Blacklight main page says: "Essentially all power sources: thermal, electrical, marine, rail, aviation, aerospace, as well as, automotive sources become untethered from an electrical distribution or fuel infrastructure and are also independent of the sun, wind, or other external variable power sources at capital cost of 1 to 10% that of historic systems." In other words, and just as Eugene Mallove said in his Open Letter, we would have a *world of abundant, clean, and safe energy from sources that have no centralized geopolitical control.* Not only is this new energy abundant, clean and safe but it is VERY cheap by historical standards, having a capital cost of 1 to 10% that of previous systems.

The conventional energy choices we have been focusing on are key components of the growing litany of geopolitical, environmental and economic nightmares all across the globe. The developments in new energy as discussed here are proof positive that we have it within our reach to end these nightmares. Just a little extra shove on our part is all that is needed.

6

HEEDING NATURE'S ECONOMIC SYSTEM

Fears over climate change have set into warp speed motion a mushrooming pileup of confusing and conflicting public policies, corporate goals and individual choices that rarely if ever deal head-on with land use as a major factor in climate change, much less address in any meaningful way the increasingly urgent environmental, economic, political and moral questions concerning land use.

As a result, energy solutions are limited to what can be "smart" metered and justified in terms of CO2 emissions, and agricultural products are limited to what can be mass-produced, and/or used as carbon sinks, no matter how much life-dependent soil carbon is destroyed in the process. At the same time, climate remediation and weather modification experiments are given the green light by policy makers and academic and corporate allies, even as the public is effectively kept in the dark about these projects, and their environmental effects.

Conveniently, the official anointing of CO2 as the world's biggest bogeyman has tidied up the means by which pollution externalities can be measured and traded. As such, CO2 has become another commodity which can be used as a lucrative profit center via the cap and trade markets, thereby facilitating the privatization of our atmosphere and the world's remaining rain forests.

As mentioned in the last chapter, a major portion of the so-called carbon markets, once they get past the current slump, will center around hugely profitable, hugely risky derivatives. This despite warnings from some economists that "speculators will end up controlling U.S. carbon prices, and their participation could trigger the same type of boom-and-bust cycles that have buffeted other commodities." (1) It is within these boom and bust cycles that the worlds of high finance and economics meet.

Economics has been defined as the study of resource allocation under scarcity, a definition which almost begs pejorative references to it as the dismal science. And because virtually all resource allocation involves land use, it is both natural and logical that the introduction of agriculture and its commodity-style food system would lead to an economic treatment of land.

So it is that land used by agriculture, along with land used for other extractive industries, has played a key role in both the analyses of, and writings on, the economy ever since ancient and medieval times. During most of this time, "land was used as an inclusive term for the natural environment, covering entities such as oceans, atmosphere or solar energy." (2)

Land Gets Redefined

The ushering in of the Industrial Revolution, however, was accompanied by a narrowing perception of land and the natural world's contribution to man's well-being. A new strain of economic thinking, generally classified as neo-classical, evolved around this narrowing perception. (3) Whereas labor, capital and land inclusive of the natural environment had once been considered the three primary inputs for gauging economic outcomes, land now became merged with capital.

One reason – and ultimate consequence – for this evolving view, as explained by one award-winning economics journalist,

> "was that the classical world view implies diminishing returns. Since the supply of land was fixed, it would become ever scarcer. . . . But . . . [t]echnical progress seemed to offset any tendency towards diminishing returns. So assuming that land, like capital, could be effectively expanded, without limit, via land-augmenting technical progress, seemed to be the right thing to do. . .
>
> The idea that land and capital are the same thing is evidently ludicrous. It requires us to believe that the economic machine is self-sustaining — a sort of perpetual motion machine. . .
>
> Yet there would be no economy — indeed no humanity — without a constant inflow of natural resources into the system: what lies above our heads (the sun and the atmosphere), what lies close to us (the soil, the seas and location itself) and what lies beneath us (fossil fuels, metals and minerals and heat).
>
> Humanity does not make these things; it exploits them. Some of these resources are also appropriable [appropriated] and so a source of unearned personal wealth." (4)

It was not until the late 19th century that American journalist, social philosopher and political economist Henry George correctly identified misallocation of land, in its most inclusive sense, as a major element missing from economic thought. His book, *Progress and Poverty*, published in 1879, became an immediate best seller with millions of copies sold worldwide in dozens of languages. It has remained an all-time best seller on economic theory and policy.

As a serious observer of the manner and speed with which the Industrial Revolution was then transforming both America and Europe, George began to wonder how it was that such unprecedented increases in productivity seemed to create poverty and want, instead of wealth and abundance. To George, this strange paradox made it seem "as though a great wedge were being forced, not underneath society, but through society. Those who are above the point of separation are elevated, but those who are below are crushed down."

After examining extant scholarship and opinion on the subject, George rejected outright the Malthusian-based idea that population growth inevitably led to poverty because of the excess drain on what British political economists called the "wage fund." Separating himself from these old-school classical economists who viewed labor, land and capital in terms of workers (whose income was derived from wages), landed aristocracy (whose income was derived from rents), and capitalists (whose income was derived from interest), George determined that land, labor and capital were instead functional categories distinguishable according to the conditions under which each were made available for production.

Thus land was to include the entire material universe, except for human beings and their products. Labor was to include all physical and mental human exertion that was involved in the production of wealth. Capital was to include the products of labor which are used in the process of production. This more nuanced approach introduced the idea that although land could indeed *not* increase, both labor and capital could. Rather than merging land with capital, as the newly emerging neo-classical economists were wont to do, George maintained that land needed to be treated separately from capital, in accordance with the conditions under which land was made available for economic production.

Land, in its more inclusive sense, was not to be confused with capital. While land can be *exchanged* for capital and while land can be used to produce income and goods – land is not itself capital. In addition and while land might be re-usable, it is nevertheless finite. Conversely, capital could expand even when much of it became "fixed" in land.

George also observed that land – again in its more inclusive sense – is subject to monopoly ownership, generally acquired as a result of boom and bust cycles. Monopoly ownership of land and its resources is of course a zero-sum equation, because such monopolies can only be accomplished by shrinking the holdings of others.

M. Mason Gaffney, who decided to pursue a PhD in Economics at Berkeley as a result of reading Henry George's *Progress and Poverty* while still in high school, explains George's main thesis this way:

> "These twin plagues [of chronic poverty and ruinous cycles of boom and bust] arose from concentrated ownership of land, compounded by land speculation. Large landowners and speculators (often one and the same) held the best land idle or underused, forcing labor onto marginal land and driving down wages. Collapse of speculative land price bubbles caused periodic slumps.
>
> (By 'land' George meant exclusive rights to use natural resources in a specified territory. It included mining, water, fishing, and timber rights, road and rail rights-of way, and some patents. George emphasized the high value and productivity of urban land, which facilitated communication and trade. Today, we would add to "land" such items as taxi medallions, telecommunications licenses and pollution 'rights'.)" (5)

The neo-classical economists, impressed as they were by the productive wonders wrought by new land-use technologies such as the McCormick Reaper, by and large denied George's distinction between land and capital and "undertook to purge land from economese." As Gaffney later pointed out in a working paper which he titled *Land As a Distinctive Factor of Production*: "Hundreds of books on economic theory are published with 'land' absent from the index. Denial is reinforced by dominant figures using sophistical, pedantic cant, which students learn to ape to distinguish themselves from the laity and advance their careers." (6)

In a footnote to this same document, Gaffney describes the process by which this "purging" has occurred, echoing some of Eugene Mallove's own discoveries about the manner in which academia can be manipulated or otherwise bought off:

> "Careers both inside and outside academia are much influenced by 'deep lobbying,' as described by William Greider, 1992, *Who Shall Tell the People?*, pp. 42-59. 'Deep lobbying' is targeting public opinion several years in the future, by building allies in think tanks, academia, the media, and select activist groups. Greider gives as an example the effort of polluting interests to undo the Superfund law. They chose The Conservation Foundation, engineering the selection of William K. Reilly as environmental czar. . .
>
> Greider emphasizes the role of economists, et al., as hired guns. The eagerness of many college professors and administrators to get grants at any price must be experienced to be believed, but I can attest from personal observation that it drives much of the profession and its attitudes." (7)

Accounting for the Needs of the Soil

Years earlier, in 1965, Gaffney had penned a scholarly paper that took Henry George's key ideas to the next logical step. Titled *Land Rent and Soil Depletion*, the paper was "an attempt to define land rent net of soil depletion." Arguing against the prevailing economists' view that all economic values are either funds or flows, Gaffney warned that land value is not an ordinary economic fund because mankind cannot add to it nor can he draw from it as from a true fund. (8)

Gaffney framed his discussion in terms of defining new economic categories that would reflect the impact of soil depletion on the economic aspects of soil, that is to say land rent, which he defined as "an income net of soil depletion." Land rent, explained Gaffney, is analogous to the legal concept of usufruct, which involves "the right to the income of an estate without impairment of its substance." This is important he says because "depletion constitutes sale of the substance of the resource, and the corresponding payment is not income but a transfer, the liquidation or amortization of a fund, comparable to sale of title to part of the land itself."

Because "the net income accruing to the land will be higher than it would be when fertility has to be maintained" some kind of mechanism was needed whereby a farmer's income would not be consumed by his attempts to maintain what Gaffney called the finite funds of virgin soil fertility. Anticipating resistance from what he describes as soil fundamentalists, private property fundamentalists and Malthusians, Gaffney devotes the bulk of his discussion to presenting four possible economic categories for soil analysis. These are: Pure flow resource, or ground matrix; Conservable flow elements; Revolving fund of soil nutrients (supplied by man), and Expendable surplus.

While the focus of the paper was to explore methods by which to arrive at "land rent" incomes net of soil depletion, Gaffney at the same time managed to "sow some wild oats, hopefully germinable, on issues in conservation, tax policy and mineral depletion:"

> "The [ground] matrix is the perdurable [most durable] base of the soil. . .
>
> The 'conservable flow' element of virgin soil fertility is that which it takes some pains to keep in its original state, but is worth those pains because they are less than the cost of replacement and less than the present value of future income from the element. . .
>
> . . . There are man-made soils, of course, and most soils contain some 'revolving fund' elements supplied by man. *But such artificial soils and soil elements are generally out of balance with their natural environment and therefore unstable. Virgin soil is largely the product of the local matrix and climate;* these are basic determinants which work ceaselessly to reduce all alien interloping elements to the likeness of virgin soil.
>
> Artificial soils, and amendments in soils, are therefore depreciable, like structures, with time, and of course they also are vulnerable to obsolescence.
>
> *Virgin soil, then, is genuinely a unique resource, differing significantly from artifacts, even its closest substitutes. And conservation of virgin soils differs significantly from artifacts, enough so, surely, to warrant a separate entry in the dictionary of economics. . .*
>
> Returning, now, to the meaning of "rent". . . *The net rent of the conservable flow element of virgin soil is reduced by conservation costs, but there remains, except in marginal cases, a surplus over conservation cost, which*

> *surplus is part of rent [income net of soil depletion]. As such it is a gift of nature. . .*" [italics mine]

Gaffney continues his discussion by exploring various aspects of his "revolving fund" concept as well as the crucial importance of humus, itself being an essentially finite product of the local matrix and climate:

> "Now let us turn to the third element of virgin soil fertility, the "revolving fund." It is that element which is not economical to conserve, but is economical to replace or renew with materials imported from off site. It is like an inventory. Examples of revolving funds are the scarce nutrients nitrogen, potassium, phosphorus, and calcium, which arc removed from the site, embodied in crops and livestock, or leached off in the productive process, then replaced commercially. . .
>
> Revolving funds are largely a feature of commercial agriculture, in which crops are exported from their sites so that nutrients are not entirely replaced by the closed cycles of natural conditions, and so must be imported. Closed natural cycles are not "revolving" of natural funds as the phrase is used here. . .
>
> *Humus is derived from surface organic matter. Minerals circulate from humus to crops to humus. But the circulation is vertical, that is on-site, as opposed to the horizontal or off site circulation of revolving funds. Why is that distinction important? Because the humus circulation is self-sustaining. . .*
>
> *. . . Humus feeds the crops that then renew it, in a continuous integral process.* Humus cannot usually be withdrawn and then replaced, economically: without

> humus, poor crops; with poor crops, little humus-increment. To a degree, humus reserves can be drawn down for later recharge. But on the whole, the mechanism must be kept in continuous operation, for liquidation is not reversible except at a much advanced cost. Those are traits of what we have called the conservable flow elements of virgin soil, not the revolving fund. . ." (9) [italics mine]

Within the context of Gaffney's discussion, it is useful to recall that it is the ranchers of Zimbabwe and original thinkers like Joel Salatin and others involved in regenerative forms of agriculture, as well as researchers like those at the Soil Carbon Institute and Sawsomme, that are redefining what it means to go organic, by devising mechanisms through which humus reserves can be continuously renewed while human needs are served using only a portion of the surpluses of nature's bounty. Some of these individuals and groups we met in Chapter 3, where we learned that we actually do now have the knowledge and experience (as well as the urgent need) to put in motion those kinds of agricultural practices that are required to heal the earth while at the same time provide us the ability to feed the world's billions *20 to 30 times over* with truly nourishing, biologically required, nutrient dense foods.

Wendell Berry's *The Gift of Good Land* suggests exactly why the kinds of practices described in Chapter 3 and the land rent discussion put forth by Mason Gaffney would provide a more complete approach to organic farming than some one-size-fits-all, legally defined "organic" farming system: "An organic farm, properly speaking, is not one that uses certain methods and substances and avoids others; it is a farm whose structure is formed in imitation of the structure of a natural system that has the integrity, the independence and the benign dependence of an organism." (10)

Management techniques must, in other words, be determined by a farm's unique locale and environment, rather than some one-size-fits all prescription which might, ostensibly at least, provide for efficient measurement of organic soil carbon. In short, regenerative forms of agriculture are unavoidably unique and site-specific and it is only through such management that carbon sequestration would become a natural and much-needed by-product of maintaining soil health and fertility.

Yet, as the reader may recall from earlier chapters, a variety of industrial-style agricultural and mining practices have for millennia been destroying the very structure of our soils, of which humus and soil carbon are key elements. These practices, which include but are not limited to land clearance for the purpose of monocrop farming, chemical fertilization, denuding, "pest" removal and irrigation, have led to "a greater than 50 percent decline in native humus levels over the years on many of the farms throughout the U.S. [Moreover] this represents a loss that probably will never be recovered." (11) Perennial agriculture, also known as carbon-sequestering agriculture and regenerative agriculture, helps solve multiple problems, including the all-important one of rebuilding humus.

As Gaffney alludes, humus is a crucially important element in soil fertility, particularly in terms of mineral availability, as well as soil structure. It is renewable, but only to a point, and only with proper care. In the starkest of terms,

> "Humus in the soil has more real value than money, real estate, stocks or bonds. Its value doesn't fluctuate; it doesn't become scarce in a recession; it's worth can't be depleted by inflation and it can't be stolen. It is the direct or indirect source of sustenance for all life on earth. It can sometimes be lost by environmental changes but more often, its demise is from either the

> apathy or the inadvertent errors of the steward who tends it.
>
> Humus is a renewable resource. Its presence in the soil can be maintained indefinitely. Unfortunately, many agricultural and horticultural practices are essentially mining humus. Like other mined products, such as coal, minerals and oil, the natural resource can eventually be exhausted." (12)

As we have discussed, the full cost of continual withdrawals from the finite fund of soil fertility, which Gaffney calls conservable flow and which represents a permanent social wealth, is intimately tied to income (and real wealth) that, in the final analysis, can only be derived from the land. Our collective failure to understand this fact is a primary cause of war, which itself destroys the finite fund of soil fertility. Thus, as perceptively observed by Charles Walters, himself a trained economist, monetary reformer, soil scientist, author and founder of ACRES USA in his 1971 classic *Unforgiven: The Story of How America Has Exchanged Parity Agriculture for Parity War* :

> "The economic law that requires permanent wealth to produce wealth that is destroyed quickly in consumption (as food or oil is) is the key. It is exactly this law that requires the public works project called war. In war, as in pyramid building and space probes, the more permanent sort of wealth is destroyed, and takes up the economic role once held by parity food. Although each of these destructive activities remains a dwarf when compared with the constructive activity of feeding the people, it is true that *the nation that degrades either the production or the income of its agriculture thereby condemns itself to war."* (13) [italics mine]

Over a century and a half earlier, in an 1816 letter to Wm. H. Crawford, Thomas Jefferson added another dimension to Walters" observation concerning the need to preserve the finite fund of soil fertility via "parity' food through his own explanation of how debt reliant monetary systems promote both speculative activity and wars of aggression. In so doing he almost eerily predicted today's multiple crises:

> "No earthly consideration could induce my consent to contract such a debt as England has by her wars for commerce, to reduce our citizens by taxes to such wretchedness, as that laboring sixteen of the twenty-four hours, they are still unable to afford themselves bread, or barely to earn as much oatmeal or potatoes as will keep soul and body together. And all this to feed the avidity of a few millionary merchants and to keep up one thousand ships of war for the protection of their commercial speculations."

In 1935, more than a century after Jefferson penned those words, and during the time when Charles Walters was just coming of age, Major General Smedley Butler, one of the most decorated Marines in U.S. history, gave an astonishingly forthright speech which provides a unique personal perspective, told from a soldier's point of view, on the economic role of war. This speech fleshes out Jefferson's "avidity of a few millionary merchants" even as it demonstrates how war can "take up the economic role once held by parity food" via the monetary system. The speech later formed the basis of Butler's book *War Is A Racket*. Here is an excerpt:

> "War is just a racket. A racket is best described, I believe, as something that is not what it seems to the majority of people. Only a small inside group knows what it is about. It is conducted for the benefit of the very few at the expense of the masses.

I believe in adequate defense at the coastline and nothing else. If a nation comes over here to fight, then we'll fight. The trouble with America is that when the dollar only earns 6 percent over here, then it gets restless and goes overseas to get 100 percent. Then the flag follows the dollar and the soldiers follow the flag.

I wouldn't go to war again as I have done to protect some lousy investment of the bankers. There are only two things we should fight for. One is the defense of our homes and the other is the Bill of Rights. War for any other reason is simply a racket.

There isn't a trick in the racketeering bag that the military gang is blind to. It has its "finger men" to point out enemies, its "muscle men" to destroy enemies, its "brain men" to plan war preparations, and a "Big Boss" Super-Nationalistic-Capitalism.

It may seem odd for me, a military man to adopt such a comparison. Truthfulness compels me to. I spent thirty-three years and four months in active military service as a member of this country's most agile military force, the Marine Corps. I served in all commissioned ranks from Second Lieutenant to Major-General. And during that period, I spent most of my time being a high class muscle-man for Big Business, for Wall Street and for the Bankers. In short, I was a racketeer, a gangster for capitalism.

I suspected I was just part of a racket at the time. Now I am sure of it. Like all the members of the military profession, I never had a thought of my own until I left the service. My mental faculties remained in suspended animation while I obeyed the orders of

> higher-ups. This is typical with everyone in the military service.
>
> I helped make Mexico, especially Tampico, safe for American oil interests in 1914. I helped make Haiti and Cuba a decent place for the National City Bank boys to collect revenues in. I helped in the raping of half a dozen Central American republics for the benefits of Wall Street. The record of racketeering is long. I helped purify Nicaragua for the international banking house of Brown Brothers in 1909-1912. I brought light to the Dominican Republic for American sugar interests in 1916. In China I helped to see to it that Standard Oil went its way unmolested.
>
> During those years, I had, as the boys in the back room would say, a swell racket. Looking back on it, I feel that I could have given Al Capone a few hints. The best he could do was to operate his racket in three districts. I operated on three continents."

The reader might observe that today we have exchanged the previous but arguably still extant trifecta of war, land use and monetary policy, as described by the above quotes, with climate change, land use and monetary policy. In both cases, monetary policy is the most important, and least understood, element of the trifecta in question.

The Importance of 100% Reserves

Unlike Gaffney and Walters, Viktor Schauberger was not an economist, but as was discussed in Chapter 5, he was a highly respected ecological pioneer and "new energy" inventor. Like Gaffney and Walters, Schauberger was also a serious student of agriculture. As such, Schauberger was deeply concerned about what Gaffney called virgin soil fertility and its "finite funds" – or what he himself called Nature's base capital.

Unlike Gaffney and most economists, Schauberger perceived two distinct economic/monetary systems, one being what he called the techno-mechanical system and the other being what he called Nature's Economic System, in which 100% of Nature's base capital – that is, its mineral resources, the replenishment and bioavailability of which is dependent upon carbon-rich humus – must be held in untouchable reserves.

Schauberger understood the importance of not just maintaining Nature's base capital but also of allowing Nature to increase her capital by a small percentage each year so as *"to allow for growth, movement and evolution of new life forms."* [italics mine] Taking Gaffney's land-rent-net-of-soil-depletion proposal a step further, Schauberger believed it was man's sacred responsibility to find ways to live sustainably off Nature's surpluses, not her base capital, which surpluses, like Gaffney, he described as a gift of nature.

In Schauberger's techno-mechanical economic system – which is what we still operate under today – the opposite occurs. This is to say, we are continually drawing down Nature's reserves by investing in highly inefficient and hugely wasteful industrial-style technologies, including energy and agricultural technologies, which themselves either progressively draw down nature's finite reserves – requiring constant and substandard "revolving fund" replacement – or which destroy the "ground matrix" altogether. (14)

Although Charles Walters framed the land issue in terms of parity agriculture (farm income net of production costs) as opposed to land rents and although he believed – correctly – that it was not distant suitcase farmer/investors but small landholder farmers who were the most acutely aware of the need to preserve nature's "base capital," he would most assuredly have agreed with both Schauberger and

Gaffney that society's focus should be on encouraging the conditions whereby the returns to small farmers would reflect the true value of farm products, that is to say income net of soil depletion and production costs.

Along these exact same lines, Thomas Jefferson, in 1813, wrote in a letter to his son-in-law John Eppes, then in Congress: "The adequate price of a thing depends on the capital and labor necessary to produce it. (In the term *capital*, I mean to include science, because capital as well as labor has been employed to acquire it.)"

Both Jefferson and Walters understood what Mason Gaffney and other economists so often miss and what Schauberger brushed up against in his two perspectives on economic systems – that "true rent" (which is to say income net of soil depletion and production costs) was wholly dependent on adoption of the right monetary system, as prescribed by Article I, Section 8 of the United States Constitution: "The Congress shall have the Power. . . To coin Money [and] regulate the value thereof. . ."

It is within this document that we find, as Walters reminded us in his book *Unforgiven,* that "our forefathers took steps to protect the economic freedom of the United States by giving to Congress, elected by the people, the power to provide a monetary system independent of the monetary systems of other countries, and to regulate the value of the dollar, adopted as our monetary unit, or measure of value. This power automatically gave Congress the right to determine the value of the United States production in terms of United States money." (15)

Unfortunately and although it had originally been met with intense and widespread opposition, Alexander Hamilton's financial plan suborned Congress's power to determine the value of U.S. production in terms of U.S. money. Intended to quickly establish world wide recognition of the nation's credit, the plan centered on the creation of the nation's first central bank, known as the Bank of the United States, which served as a forerunner to the present day Federal Reserve Banking System. The majority of stock was owned by private investors, with the government owning a minority share. These private investors included large numbers of foreign investors. For instance, out of the original 25,000 shares issued, some 18,000 were purchased by English investors. The advantage here, in Hamilton's eyes, was that the Bank of England would then provide loans to the United States in exchange for securities of the United States.

Similar to the Federal Reserve, Hamilton's plan provided that the Bank of the United States, and not the U.S. government, was to create the nation's currency. This currency was to be created out of debt, as a result of bond sales and loans made to large merchants and manufacturers, as well as prominent politicians and the U.S. government, using a reserve ratio of roughly 33 1/3%. Taxes would be used to pay only the interest on the debt which ideally would remain intact so said debt could both serve the purpose of money and bind the people, especially bond investors, to the government. In addition, regular payments on the interest on the debt would simultaneously make the U.S. the best credit risk in the world and create an attractive system of bond investments for America's wealthier citizens. These bonds would not only further bind investors to the government but they, together with loans, could form the nation's money supply.

One of Hamilton's primary goals was to provide the means by which to quickly establish an American manufacturing base that could rival that of Europe and England, eventually turning the United States into a powerful fiscal-military state. In order to jump start large manufactures in the United States, Hamilton proposed the use of subsidies for select industries, paid for through a program of tariffs and excise taxes. The yet-to-be formed Jeffersonians rejected outright Hamilton's use of subsidies as an economic stimulus tool, charging that subsidies unnecessarily and unwisely extended the authority and influence of government. The tariff question was somewhat more nuanced, with the embryonic Hamiltonians viewing tariffs and excise taxes more as sources of revenue and a means by which to grow a large manufacturing base rather than as a mechanism by which to protect local producers, including farmers, thereby providing a mechanism by which to determine the value of U.S. production in terms of U.S. money.

Similar to the Bank and subsidy proposals, the tariff program as first established by the 1789 Tariff Bill was hotly contested by the nascent Jeffersonians, then under the leadership of James Madison, who supported "discriminatory" tariffs and quotas on imports. The weakness of the Articles of Confederation had already, prior to the adoption of the Constitution, prompted no less than 10 colonies to place duties and "discriminatory" tariffs on British goods. Thus, many of the nation's small manufacturers and tradesmen, who had been petitioning the newly formed Congress for relief from the flood of imported foreign goods, especially English goods, themselves favored the Jeffersonian (or as it was then the Madisonian) plan.

The Tariff Bill of 1789 was the first major bill to be considered by the new Congress and the first to be passed, but only after James Madison's attempts to "discriminate" against British imports had been dropped as part of a compromise with the Hamiltonians. Asserting that those items subject to high duties under the Tariff Bill "were pretty generally taxed for the benefit of the manufacturing part of the northern community," Madison sought to modify the terms, this in an attempt to balance sectional conflicts. Despite this effort, he had to concede that the South, which was then the main wealth-producing part of the nation, would be forced to "shoulder a disproportionate share of the financial burden involved in the transforming the United States into a commercial, manufacturing, and maritime power."

The more nuanced approach of the Jeffersonians was calculated to provide not just revenue or the means by which to grow large manufactures but also as the means by which to regulate trade in order to both protect existing American industries *and* to grow the nation's wealth through its own agricultural, merchant and smaller scaled manufacturing base. Thus, restriction on imports of items already domestically produced, or able to be produced, in amounts sufficient to meet demand meant that prices would not be forced to an artificially low level by flooding the market with unneeded imports. Similarly, placing a tax on specific, "overly cheap" imports would force the prices of those imports up, thereby encouraging consumers to buy domestic products, at prices which offered producers profits sufficient to cover costs while providing a reasonable income. Essentially, the Jeffersonians sought methods by which to determine the value of U.S. production in terms of U.S. money, not in terms of building a world super power via a debt-based monetary system.

Many at the time were fully convinced that instead of securing a nation of merchants, small manufacturers and farmers, Hamilton's deliberate use of the bank as an "engine of influence" was creating a class of men that would enrich themselves through stockjobbing, speculation and similar financial manipulations. In point of fact, Hamilton's plan clearly favored prominent members of the business, commercial and political classes – and helped lay the groundwork for Wall Street to become the financial hub of the world. Perhaps even more importantly, it firmly established the practice of creating the nation's currency out of debt which remains to this day.

In direct contrast to the 200 year plus practice of creating the nation's currency out of debt, the late Irving Fisher, together with a handful of similarly discerning economists, adopted views similar to Charles Walters' as these related to Constitutional money. As Professor Emeritus of Economics at Yale and one of America's greatest mathematical economists, Fisher developed monetary views that also ran parallel to Schauberger's two distinct economic systems.

Fisher, in fact, had been part of a prominent group of economists who drew up recommendations for a series of banking reforms which became known as the Chicago Plan, major elements of which later resurfaced as "A Program for Monetary Reform," issued in 1939. (16) Over 230 economists from 150 universities approved of the 1939 plan without reservations, and another 40 supported it with some reservations. Among other things, the plan called for a national currency that would be issued and regulated by our government, as the Constitution dictates, and it called for a 100% reserve requirement for banks, thereby ending the fractional reserve system. Under a 100% reserve system, banks could only lend the money they possessed.

Despite the widespread recognition achieved by the Chicago Plan, our current monetary system of course still relies on the banking system to create the nation's currency out of debt by lending money they don't actually have via fractional reserve deposit expansion. Banks literally create what serves as money out of the loans they make. The amount of what Jefferson referred to as "nominal money" and "watered down paper" created as debt by the banks is governed, theoretically at least, by the reserve requirement, also known as the money multiplier. Using an example from *Investopedia,* we can see just how the "money multiplier" works:

> "We start with the amount banks initially take in through deposits and divide this by the reserve ratio. If, for example, the reserve requirement is 20%, for every $100 a customer deposits into a bank, $20 must be kept in reserve. However, the remaining $80 can be loaned out to other bank customers. This $80 is then deposited by these customers into another bank, which in turn must also keep 20%, or $16, in reserve but can lend out the remaining $64. This cycle continues – as more people deposit money and more banks continue

> lending it – until finally the $100 initially deposited creates a total of $500 ($100 / 0.2) in deposits." (17)

Apart from the fact that the bank is in effect lending out money based on what its customers have on deposit, and not what it actually has in its "vaults" – thereby risking customer deposits – we can also deduce the following with respect to the multiplier effect, this the result of "double entry accounting":

> "A lender starts with a million dollars and the borrower starts with zero. Upon extending the loan, the borrower possesses the million dollars, yet the lender feels that he still owns the million dollars that he lent out. If anyone asks the lender what he is worth, he says, 'a million dollars,' and shows the note to prove it. Because of this conviction, there is, in the minds of the debtor and the creditor combined, two million dollars worth of value where before there was only one." (18)

Irving Fisher's own careful study into the causes of economic depressions led him to the conclusion that the fractional reserve deposit expansion monetary system was to blame for recurring boom and bust cycles. This conclusion, said Fisher, was not new. It had in fact already been arrived at independently by many writers over the previous 100+ years. Fisher further demonstrated that economic depressions are always caused by the accumulation of too much debt, and/or too great a contraction of the circulating medium, which has for nearly all of our history been debt serving as money.

While commenting on the negative effects of the fractional reserve deposit expansion monetary system (via the banking system) such as ours is today and which is the basis for debt serving as money, Fisher in 1936 pointed out that "few people not versed in banking ever dream that their money 'in the bank' is not in the bank; and few people versed in banking ever dream that the money which is not there ought to be there, or that it is practical to have it there." (19) (20)

Similar to Schauberger, Fisher and his colleagues maintained that the monetary (or economic) system was to blame for recurring boom and bust cycles, because it allows banks to simply "manufacture the money which they lend" through the fractional reserve system, creating loans (and money-as-debt) from non-existent "base capital."

Thomas Jefferson was but one of our founders who advocated Fisher's 100% money solution, as demonstrated in an 1813 letter to his son-in-law John Eppes: "No one has a natural right to the trade of a money-lender, but he who has the money to lend" and other statements. As revealed in various letters from this time period, when the first Bank of the United States was up for re-charter, Jefferson himself was urging the creation of interest-bearing Treasury notes as a means by which to get Americans once again used to the idea of government issued money.

Though he well understood that a sizable portion of such notes were likely to be hoarded by those who possessed them, his letters indicate that he believed that these "hoarders" might at least partially function as unofficial 100% reserve "mini-banks" thereby loosening the hold of the fractional reserve banking system. At the same time, he, like others, maintained that non-interest bearing Treasury Notes "bottomed on a tax that would redeem them" were all that was needed to satisfy the currency needs of the nation. It should be understood that these were not entirely new ideas, precedent having been set by the colonies before the Revolution for both types of currencies.

Ecological economist Herman Daly, successor to Irving Fisher and his allies, examined how a "100% Money" system could create what he called a "steady state economy" in an article written in early 2010. Among these benefits are the following:

> "First . . . it would restrict borrowing for new investment to existing savings, greatly reducing speculative growth ventures—for example the leveraging of stock purchases with huge amounts of borrowed money would be severely limited.
>
> Second, the fact that money no longer has to grow to pay back the principal plus the interest required by the loan responsible for the money's very existence lowers the general pressure to grow. Money becomes neutral with respect to growth rather than biasing the system toward growth.
>
> Third, the financial sector will no longer be able to capture such a large share of the nation's wealth, leaving more available for meeting the needs of the poor. A steady state economy is not viable if it means a steady state of poverty for any significant proportion of the population.
>
> Fourth, the money supply would no longer expand during a boom, when banks like to loan lots of money, and contract during a recession, when banks try to collect outstanding debts, thereby reinforcing the cyclical tendency of the economy. Reducing the risk of recession reduces the need to accumulate more to get us through the bad times." (21)

Henry George had ascertained that the twin plagues of chronic poverty and ruinous cycles of boom and bust arose from concentrated ownership of land. The alert reader will see from Daly's description above that both boom and bust cycles and concentration of wealth in the financial sector (not to mention excessive and unwarranted growth arising from speculative activity) will be minimized if not eliminated altogether with a 100% money system, or as Daly described it "a steady state economy."

Economist Mason Gaffney, drawing on George's discovery that land was often the central vortex around which boom and bust cycles revolved, makes note of the tendency of banks to lend freely on land during land booms. While not coming to the same conclusion of monetary reform as Daly does in the above passage, Gaffney accurately describes another disagreeable aspect inherent within our debt reliant monetary system, particularly as it pertains to speculative-induced "degeneration of enterprise":

> "During a land boom, financial institutions lend freely on land. After a while, mortgages secured by real estate (either directly or through the corporate veil) become the major asset of banks. Credit follows collateral, and then helps boost its collateral value in a positive feedback loop. In periods of high and rising land prices, borrowers get used to pledging land to secure loans, and lenders get used to demanding it.
>
> The credit is often used to buy still more land, to reserve for possible future use and at the same time to withhold from competitors. Such concentration and market control form the ugly side of extant western capitalism, when enterprise degenerates into greed and acquisitions supplant innovations." (22)

In his own book titled *Unforgiven: the Story of How America Has Exchanged Parity Agriculture for Parity War*, published in 1971, Charles Walters colorfully and succinctly traces the social and monetary roots of the moral degeneration which Gaffney attributes to western capitalism and which is in many ways today coming full circle back to its origins:

> "Once upon a time the nobles of Europe believed they had accomplished the physical impossibility of perpetual interest simply by owning all the land and raking in tribute from the peasantry, and when death finally pulled down an economic maggot, then there was always the heir. The historical showdown arrived, of course, and *the 'new nobles' were forced to invent a more subtle form of tribute taking. It came on as debt, interest, compound interest, and all the institutional arrangements required to make the producing community share its income with the creditor.* For centuries the details have stacked up, but economists [by and large] have failed to draw the appropriate conclusions." (23) [italics mine]

As Walter's knew, the old nobles of Europe were beneficiaries of the debt reliant monetary system which had originated in ancient Mesopotamia. As esteemed economist and Mesopotamian scholar Michael Hudson tells us, "It was in this epoch, over two thousand years prior to classical antiquity, that the basic elements of modern economic relations first appear." And it was during this era that a rigidly hierarchical economic, social and political structure began to take legal shape, starting with the peasants who were provided with a portion of land, along with tools, livestock and water for irrigation, so that they might grow food for State workers and dignitaries.

A portion of what the peasants harvested had to be paid to the State as rent. When the harvest was poor they went into debt. Since the peasantry might also own or aspire to owning their own land, they were easy targets for private loans granted by high-ranking officials and wealthy individuals who were themselves eager to get rich and seize peasant property via default. Peasants who could not pay their debt could even be reduced to debt-slavery, sometimes taking other members of their family with them into debt servitude. As overall debt began to exceed peasant ability to pay, default rates would escalate and property ownership would begin to concentrate in the hands of the creditors. Social unrest would then escalate, a situation which the reader might note is occurring today.

To solve the problem, authorities periodically cancelled all debt and restored peasants' rights, thereby interrupting the inexorable and inevitable tendency of all debt reliant monetary systems toward concentration of property ownership. These debt cancellations (or Jubilees) were done primarily to maintain peace and social stability. Inasmuch as they were specifically intended to preserve the pre-existing social order they could hardly be counted as a form of social emancipation. Moreover, they did nothing to halt the *need* for debt cancellations in the first place. Nevertheless debt cancellations did help maintain social cohesion by inhibiting the development of large private domains and by ensuring that peasants would have direct access to land.

It was from this last provision that the legal concept of usufruct evolved, with the first examples being found in the Law of Moses and the Hammurabi Code. The Law of Moses actually prohibited property owners from harvesting the edges of their fields, so that the poor might glean their food from these areas.

By preserving the social and economic order and limiting the rise of inequality through periodic debt cancellation, authorities could continue business as usual, overseeing the construction of temples, state houses and irrigation systems along with other public projects deemed worthy or necessary, including war for new land. Many of these projects, including war, were achieved through recurring waves of debt expansion on the backs of the peasants – and the land.

Debt cancellations ceased in Mesopotamia around 1400 BC, but the practice spread to Egypt, ancient Greece and Rome, where such cancellations increasingly were prompted by social upheaval. (24) Some governments, such as Athens and early Rome, even installed their own debt-free currency which lasted for hundreds of years but the private debt/money powers seem to have always found ways to eventually reinsert themselves.

Over the following centuries, inequality increased, debt enslavement became commonplace and more and more land began to be taken taken over by big private land owners. By the end of the first century AD, the Roman Empire – by then one of the largest empires in history with contiguous territories throughout Europe, North Africa and the Middle East – had a tiny minority of super wealthy people, 2000 of whom owned all of Europe. This in an empire numbering 50 million. War and slavery were two of its chief characteristics. (25)

Usufruct, which involves the legal right to use or profit from something belonging to another, became a key feature of the Roman-based legal system. As the online *Encyclopedia Britannica* explains, one of "its most important applications was in the determination of the property interests between a slave held under a usus fructus (Latin: "use and enjoyment") bond and a temporary master. Any property acquired by a slave as a result of his labor legally belonged to the master." (The reader might consider the similarities to modern day business practices, particularly with respect to intellectual property rights.)

The old nobles of Europe, or the landed aristocracy, were still firmly entrenched by the time of the American Revolution, protected by inheritance laws passed down over many centuries. It is of course arguably true that new and bolder experiments in debt reliant economies helped grease the wheels of commerce and spur all manner of artistic endeavors and technological innovation, especially from the Renaissance forward. It is also true however that many of the wonders of the Renaissance and after came not from any sweeping social, economic or political reform but rather from the largesse of wealthy individuals such as the Medici family, who accumulated their extraordinary wealth and power through their innovations in banking.

Following a familiar pattern established in ancient Mesopotamia, this new and "improved" debt-driven economic system also brought out the worst in human nature. Evidence abounds but well known examples include the early 18th century South Sea Bubble with which the founders themselves were intimately acquainted – and the brutal factory system that emerged in England in the 1750's at the beginning of the Industrial Revolution, spreading quickly thereafter to Europe. It was this specific type of manufacturing that was so vehemently and vociferously reviled by Thomas Jefferson, having taken it upon himself to witness its effects up close and personal during his time in France in the 1780's.

It was also Thomas Jefferson who, while serving in the legislature of the Commonwealth of Virginia prior to the Revolutionary War, introduced a lengthy series of land reform proposals aimed at dismantling Old Europe's inheritance laws which protected the landed aristocracy. His proposals included the elimination of the practice of granting still more acres to individuals who already had acreage in abundance and at the same time recommended that small, unencumbered, 50 acre parcels be provided to the landless. His overarching, lifelong goal was to set an aristocratic-leaning society "in the current of democracy by broadening its economic base."

However, and although Jefferson most assuredly took the lead in his own large and influential Commonwealth of Virginia, he was not alone in his efforts at broadening the economic base through land reform. These efforts were "part and parcel of a movement which swept through the other states and resulted in the abolition of primogeniture and entail in practically all of them within ten years of the signing of the Declaration of Independence." (26)

7

THE EARTH BELONGS TO THE LIVING

Most, if not all, of our founders were well acquainted with the "perpetual interest" system devised by the old nobles of Europe, referred to by Walters in the last chapter. Many also well understood the "more subtle form of tribute taking" anchored in debt which required the producing community to share its income with the creditor, but few thought to relate this to the legal concept of protecting the usufruct.

Thomas Jefferson, however, expounded at length on the subject in his famous "earth belongs to the living" letter to James Madison, foreshadowing many of the same issues and ideas revisited by Walters, Schauberger, Gaffney, Fisher, Daly and others. Written while still in Paris on September 6, 1789, the letter is essentially a position piece turning on the rights of the current generation to encumber future generations. It reads in part as follows:

> "The question Whether one generation of men has a right to bind another, seems never to have been started either on this or our side of the water. Yet *it is a question of such consequences as not only to merit decision, but place also, among the fundamental principles of every government*. . . I set out on this ground which I suppose to be self-evident, *that the earth belongs in usufruct to the*

living, that the dead have neither powers nor rights over it.

The portion occupied by an individual ceases to be his when himself ceases to be, and reverts to society. If the society has formed no rules for the appropriation of its lands in severalty, it will be taken by the first occupants. These will generally be the wife and children of the decedent. If they [society] have formed rules of appropriation, those rules may give it to the wife and children, or to some one of them, or to the legatee of the deceased. So they may give it to his creditor. . .

But a material difference must be noted, between the succession of an individual and that of a whole generation. Individuals are parts only of a society, subject to the laws of a whole. These laws may appropriate the portion of land occupied by a decedent, to his creditor, rather than to any other, or to his child, on condition he satisfies the creditor. But when a whole generation, that is, the whole society, dies, as in the case we have supposed, and another generation or society succeeds, this forms a whole, and there is no superior who can give their territory to a third society, who may have lent money to their predecessors; beyond their faculties of paying. . .

[The present generation] have the same rights over the soil on which they were produced, as the preceding generations had. They derive these rights not from their predecessors, but from nature. They, then, and their soil are, by nature, clear of the debts of their predecessors. . .

> I suppose that the received opinion, that the public debts of one generation devolve on the next, has been suggested by our seeing, habitually, in private life, that he who succeeds to lands is required to pay the debts of his predecessor; without considering that this requisition is municipal only, not moral, flowing from the will of the society, which has found it convenient to appropriate the lands of a decedent on the condition of a payment of his debts; but that between society and society, or generation and generation, there is no municipal obligation, no umpire but the law of nature." (1)

It has been said that Jefferson's argument was ahead of its time, and perhaps it was. To be sure it was a gauntlet drawn against debt driven monetary systems that have plagued mankind since antiquity. His clearly articulated and oft-repeated belief was that no generation had a moral right to bind another with financial claims that would exceed the usufructory limits of the land. This belief, in sharp contrast to Hamilton's financial system which would take shape shortly after, in essence required every generation to pay its own debts, and it was inextricably tied to his life-long conviction "that a reliance on bank notes [such as the Federal Reserve Bank Notes we have now] would create economic instability and a permanent creditor faction reliant on the tax payer, while in contrast, government paper money would prevent both evils." (2)

Although a key element of the "Earth belongs to the living" letter excerpted above is about public debt, the whole of Jefferson's argument is based on the legal concept of usufruct which, as Mason Gaffney pointed out, "involves the right to the income of an estate without impairment of its substance" and which can be traced back to ancient Mesopotamia and then more formally, to Roman civil law. Thus, as Jefferson takes pains to point out, the wanton accumulation of public debt by one generation will only put off that which must be paid by succeeding generations with the produce of the earth and human labor.

In other words, "'eating up the usufruct' means extinguishing the next generation's ability to share equitably in the benefits of a natural resource. No individual or society has authority to cause such extinction, whatever personal or collective rights they may allege."(3) Rather, "each generation has the right to inherit, undiminished, the same topsoil capital that its predecessors enjoyed." (4)

Another influential commentator on bank-credit-serving-as-money was Edward Kellogg, a self-made merchant who had been ruined by the Panic of 1837. Devoting himself thereafter to a study of international finance, Kellogg wrote highly influential treatises on money, which he came to know was a creature of law, rightly created by government alone. His daughter reorganized and published his writings posthumously under the title *A New Monetary System* which can be read online. (5)

Similar to many of Jefferson's explanations of the problems associated with what the Jeffersonians called bank paper (or simply paper), Kellogg draws a distinction between the legal value and actual value of bank paper serving as money and outlines the effects of the one upon the other in a monetary system that relies on the banking system to create money from debt, which is to say from the loans the system *as a whole* makes. His explanation, like that of Jefferson, Walters and others, helps further clear up Henry George's initial question about how it was that such unprecedented increases in productivity he was witnessing seemed to create poverty and want, instead of wealth and abundance – and it speaks to Jefferson's concern about using up the usufruct:

> "Legal value belongs to anything which represents actual value, or capital. *Its existence depends upon actual value.* The worth of things of legal value depends

> upon their capability to be exchanged for things of actual value. The following illustration shows the distinction between actual and legal value, and the dependence of the latter upon the former.
>
> The national debt of England is [in the 1830's] about $4,000,000,000. It bears interest at about an average of three percent per annum, amounting to an annual sum of $120,000,000. [Here] A hundred and twenty millions of dollars' worth of the products of labor, of actual value, must be sold annually to pay the interest; to pay the principal would require a large proportion of the wealth of the country. If the paper, *the legal value which represents and secures the debt and interest,* were collected and burned, it would not diminish the real wealth of the nation. *It would merely cause a change in the individual ownership of property."* [italics mine]

Put another way, we see from Kellogg's example that England's debt is secured by collateral that has actual value – and the creditors, or those holding the notes, draw interest income on that debt. Those bank notes, or paper, are the legal representative of said collateral. It is the interest due on those notes that conflates the actual value of the collateral and requires the "eating up of the usufruct." And, if for some reason England becomes unable or otherwise decides not to pay her debt, then the creditors will have the legal right to take ownership of the collateral, which still retains actual value, resulting in a change in ownership of the property.

Because the legal value of money (or bank notes or "paper") must "represent and secure the debt and interest," its claim on goods and services (which represent the real wealth of the country) exceeds the actual value of a nation's real wealth at a given point in time, this due to the interest which must be paid. Thus, using Kellogg's example of

England, *$120,000,000 worth of the products of labor, of actual value, must be sold annually to pay the interest.*

In order to meet the on-going and relentlessly increasing demands of interest payments, the economy, and the debt/money supply must continually expand at least sufficiently enough so as to allow for payment of interest due. The "exponential laws of increase" governing debt effectively requires the present generation to borrow increasingly from the next in order to simply pay the interest on the expanding debt, resulting over time in a progressively greater shortage of money relative to debt, which is to say there is more debt representing *claims* on collateral than there is actual goods and services available to the economy. This is the key reason why economies must grow their way out of debt if they are to survive – creating "ungodly greed" and waste in their wake.

Economist and Mesopotamian scholar Michael Hudson explains the situation from a slightly different perspective than Kellogg by critiquing some of the flaws in modern day economic thought and analysis:

> "Economies change their shape as they grow. *This shape is distorted by the inherent tendency for financial claims – bonds, bank loans and other financial securities – to grow more rapidly than the economy's ability to carry them, much less to pay them off.* The volume of such claims tends to grow by purely mathematical principles of self-expansion, independently from underlying economic trends in wealth and income, and hence from the ability of debtors to pay. . . .
>
> *Excluded from the [economists'] analysis is the finding that many debts are not repayable except by transferring ownership to creditors. This transfer changes the shape of the economy's legal and political environment, as creditors*

> *act as rentiers to subordinate labor and capital to the economy's financial dynamics.*
>
> *Orthodox academic models rarely acknowledge the problems posed by the exponential growth of debt overhead.* Such models typically make government policies appear unnecessary to cope with this problem, by focusing on the kind of world that might exist if the financial overgrowth of savings and debts did not double every decade or so, having multiplied again and again over the past century.
>
> *Babylonians recognized that while debts grew exponentially, the rest of the economy (what today is called the "real" economy) grows less rapidly. . .*
>
> *Today's economists have a problem analyzing the relationship between the debt overhead and the capacity to pay.* Academic orthodoxy holds that economies can adjust to any volume of debt, given sufficient price and income flexibility to facilitate the transfer of revenue and assets to creditors. What is not recognized is that the resulting economic polarization reduces the economy's ability to function well. In addition to missing this negative feedback (the proverbial vicious circle), modern economists tend to overlook the fact that interest-bearing debt grows according to its own exponential laws of increase. The economy rarely can keep up." (6) [italics mine]

Transfer of ownership resulting from the manner in which the "exponential laws of increase" push debt overhead beyond the economy's capacity to pay is one key mathematically-induced characteristic of our current debt-dominant monetary system. *Debt-induced* currency devaluation (as separate from official manipulation of international exchange rates) is another, requiring both a continually expanding economy and by implication, continually rising wages and prices (in the aggregate) just to pay the interest costs on the debt, irrespective of the actual cost to produce goods and services. The dollar in effect loses purchasing power over time due to the demands of "exponential laws of increase" and the trick is to try to manage monetary policy so as to allow sufficient new debt/money to enter the economy so that the interest on the debt can be paid.

Not by coincidence then has the debt overhead now permeating throughout the U.S. economy prompted new, uncharted waters in the management of monetary policy. Yet, five years after the banking crisis of 2008, the U.S. economy is still faces the very real possibility of yet another serious contraction despite massive bank bailouts, ongoing QE programs aimed at raising asset prices, job stimulus programs and other rescue measures which have been employed to prop up our debt-ridden, debt-dependent economy. Despite these gargantuan efforts to right the ship, the transfer of wealth upwards has only accelerated, with 22 million children among the 1 in every 6 Americans now living on food stamps and facing bleak job prospects, even among college graduates. This of course is, as Hudson and many others tell us, due to the fact that the real economy cannot keep up with the demands of our growing debt overhead. Even worse, the debt-induced devaluation of the dollar makes it harder for individuals, local governmental entities and private businesses to pay back their debt.

While many of us fixate on inflation as the main problem with our current monetary system, the real problem is that the process of creating a society's money out of debt results – *over time* – in an insidious and progressive devaluation of the money standard, which in our case is the dollar. This dollar devaluation, like transfer of ownership, is due to the demands of the "exponential laws of increase" governing interest-bearing debt, making it harder and harder *over time* to pay off both public and private debt.

The Populists and Greenbackers of the late nineteenth century understood this phenomenon and it is why they called for a dollar with debt-paying power – not a dollar which "must first represent and secure the interest and the debt" before paying for goods and services. The farmers in particular knew first hand that the prices (income) they received for the goods they produced never kept pace with the debt they had to pay off, because of the extra demands of the debt/money they had access to.

Thomas Jefferson was yet another of countless legions who have urged others to consider the insidious effects of the "exponential laws of increase." The following is an excerpt from yet another 1813 letter to his son-in-law concerning the effort then emerging to create a second national bank – which like the first would create a national currency through borrowing:

> "The scheme is for Congress to establish a national bank. . . .This state of things is to be fastened on us, without the power of relief, for forty or fifty years. That is to say, the eight millions of people now existing, for the sake of receiving one dollar and twenty-five cents apiece, at five per cent interest, are to subject the fifty millions of people who are to succeed them within that term, to the payment of forty-five millions of dollars, principal and interest, which will be payable in the course of the fifty years. . ."

In short, the Jeffersonians – as distinguished from the Hamiltonians – believed that the legal value of money should equal the actual value of goods and services, which represent the real wealth of a nation. The people therefore should not be provided with a currency for which succeeding generations must be forced to produce goods and services solely for payment of accumulating interest on the currency. The Jeffersonians also understood, as the passage above suggests and as their own words clearly indicate, that maintaining parity between legal value and actual value can only be achieved when money is spent into circulation by the federal government debt free (rather than loaned into circulation the way it all too often has been for centuries), all taking place within a 100% money system wherein debt in the economy is not allowed to exceed its income.

In a manner similar to Henry George, the Jeffersonians would very likely have separated themselves from the classical economists who viewed labor, land and capital in terms of workers (whose income was derived from wages), landed aristocracy (whose income was derived from rents), and capitalists (whose income was derived from interest). They surely would not have approved of a monetary system which allowed creditors to act as rentiers who sought to subordinate labor, capital *or* land to their needs. And they most certainly would never have seen fit to purge land from "economese" as the neo-classical economists were wont to do. (7)

It might even be fairly argued from their writings that the Jeffersonians would, like Henry George, have placed land, labor and capital in separate functional categories. At the very same time, they would also have demanded that income in the aggregate come from actual earnings – not from claims on wealth that would "eat up the usufruct" and enable creditors to act as rentiers. For the same reason of protecting the usufruct, they would have maintained that since the income (or capital wealth) of land, labor, and capital was ultimately derived from the land, the usufruct of said land must not only be supported by a non-debt reliant monetary system but said land, in its broadest, historical meaning, must also be protected by society's laws and customs.

Unfortunately the opposite is happening. Indeed, the trend in the opposite direction is actually accelerating thanks to today's irrationally exuberant form of globalization, which itself is driven by an interlocking system of finance and trade that is wholly dependent on leverage (or debt), scalability (or economies of size), and centralized control of production, distribution and prices. All of it rests on the unstable foundation of a money system that by its very design "eats up the usufruct" – thereby necessitating hyperbolic globalization even as it fosters monopolization.

All of it despite the fact that – as Charles Walters correctly told us in his book *Unforgiven* – "our forefathers took steps to protect the economic freedom of the United States by giving to Congress, elected by the people, the power to provide a monetary system *independent of the monetary systems of other countries, and to regulate the value of the dollar*, adopted as our monetary unit, or measure of value. *This power automatically gave Congress the right to determine the value of the United States production in terms of United States money.*"

Monetary scholar T. Cushing Daniel, in his book *High Cost of Living*, similarly informed his readers that "in order to understand what real money is, the people of the whole world should expel from their minds that there is any such thing as money. Money is a creation of an act of sovereignty and is real money, legal tender money, only within the jurisdiction of the country issuing it."

In his book *Unforgiven*, Charles Walters shares a highly relevant, very prescient excerpt about free international trade from a 1934 book by the late historian Christopher Hollis titled *The Breakdown of Money: an Historical Explanation.* In this passage, which has to do with the fact that free international trade requires a price dislocation between nations, Hollis clearly explains the operational mechanics of said trade and how it destroys our nation from within by failing to determine the value of US production in terms of US money:

> "As long as the purchasing power [of the American consumer] was adequate, the American manufacturer was indifferent to foreign markets. But with domestic purchasing power reduced, foreign markets became essential to him. And, the more that he could be

> persuaded to look abroad for this market, the easier it will be to change his whole attitude towards wages. At present he is in favor of high tariffs and high wages, for he looks on the working man as his customer. But, if he can be induced to look abroad for his markets, then wages become merely an item of costs and it is to the manufacturer's interests to reduce them as low as possible. If they are reduced – *and the odium for reducing them of course, allowed to fall on the manufacturer – then American industry becomes at once a much more profitable investment for the financier, while the foreign goods can flow into free trade America to pay the interest on foreign loans."* (8) [italics mine]

Today more than ever before in our history, the value of United States production is determined by the international marketplace, which viciously pits the world's people and their governments against each other and where poor countries with low or no labor or environmental standards and plenty of cheap labor and natural resources provide products at rock bottom prices, forcing other countries to follow in a race to the bottom. All too many of these below-actual-cost products come with a "buyer beware" label carefully hidden from the purview of the average consumer.

Professor of Economics Michel Chossudovsky, who is author of *Globalization of Poverty* and similar books, provides a vivid illustration of the manner in which wealth is extracted from local producers operating within the real economy through manipulation of today's global financial economy. As Chossudovsky explains, there was in 2009 a 95% mark-up of all goods manufactured in China which were sold in the U.S. This meant that some item a company purchased in China for $1, the company could then turn around and sell for $10 in the U.S. This maneuver in turn increased the U.S. GDP by $9.

The reader may notice from this example that rather than preserving the usufruct by maintaining a direct connection between producers and income, the usufruct is instead being destroyed in order to create fictitious paper wealth. This entire process is of course made achievable, one might even say necessary, via the current monetary system, which is itself wedded to modern day free trade dogma.

Not unlike Dr. Hudson and a growing cadre of others, Dr. Chossudovsky amply documents the rapid escalation of financial manipulation over the past few decades. He, like Hudson, also warns that, as a result of the on-going crisis begun in 2008, we are witnessing not only the most massive transfer of wealth in history but also a major transformation within the financial system itself. For example, he explains how Wall Street has begun to function in a manner very similar to the way the IMF functioned in the developing world during the 1980s. One telling indication of this recent trend is that the debt-rating agencies employed by Wall Street have become far less interested in debt per se and far more interested in whether policy makers "obey orders" by implementing austerity measures similar to the structural adjustment measures imposed by the IMF on developing countries.

We have reached a place in world history where it can be said with some accuracy that the earth no longer belongs to the living, it belongs, as Hollis's words foretold, to the financiers and debt-holders who may indeed be driving a faustian bargain with the planet. This bargain is aided and abetted, as earlier noted by economist Michael Hudson, by the easily observable fact that the increasingly accelerated transfer of property typical of debt reliant monetary systems "changes the shape of the economy's legal and political environment, as creditors act as rentiers to subordinate labor and capital to the economy's financial dynamics."

Global Free Trade Legalizes Theft of Nature and People

Much of the worldwide economic and environmental devastation we see today is sold to us under the mantra of free trade, which is, as it all too often has been, really trade *of, by* and *for* the monopolies. Today's uber-extreme version of free trade began with GATT and the WTO in the 1990s. This was despite the fact that no crisis whatsoever then existed in international commerce, particularly one that otherwise might have provided some sort of justification for the draconian measures implemented under the GATT, the WTO and all subsequent agreements. These trade agreements give new meaning to Jean-Jacques Rousseau's eighteenth century observation that "civil society is little more than the result of a conspiracy of the rich to guarantee their plunder."

By way of example, two important pacts developed through the Uruguay Round of GATT were delivered in 1995 to the WTO to enforce. These were: The TRIPS Agreement, which criminalized seed saving and seed sharing and allowed for the patenting of life forms – and The Agreement on Agriculture, which legalized the dumping of genetically engineered seeds and foods on countries even as it criminalized actions to protect local, organic food systems. In other words, the Uruguay Round of GATT in 1994 together with the creation of the WTO institutionalized and legalized corporate growth based on the theft of nature and workers. In addition, the WTO, like its successors, has explicit authority under international law to decide which local and national laws are in violation of its trade rules.

More recently and flowing from provisions in the North American Free Trade Agreement (NAFTA), huge swaths of Alberta, Canada have been taken over by tar sands mining projects. Today, Alberta contains approximately 68 square miles of toxic tailing ponds, needed to contain tar sand mining waste, alone. These tailing ponds are so huge that they are considered to be one of the largest human-made structures in the world. As one online activist group details:

"In addition to greenhouse gases, tar sands operations release large volumes of acid rain-causing sulphur dioxide and nitrogen oxides into the environment. . . Two separate research studies, one by internationally recognized water expert Dr. David Schindler and another by Environment Canada, indicates that tar sands production is releasing various toxins, including napthenic acids and polycyclic hydrocarbons, into the air. They have been found in snow, water and lake sediments as far at 100 kilometres downstream of tar sands mines and upgraders.

Mining and processing tar sands also create a toxic sludge called tailings. These tailings – which contain water, sand, silt, clay, contaminants and unrecovered hydrocarbons, including numerous toxins harmful to humans, plants and animals alike – are kept in large "ponds" created to store the waste indefinitely. Tar sands mines create as much toxic tailings every day as flows over Niagara Falls in 90 seconds. The second largest dam on earth is holding back toxic sludge in the tar sands. The only larger dam is the Three Rivers Gorge Dam in China." (9)

Canada's tar sands deposits are located under its boreal forest, which is the second largest intact forest in the world. It is also one of the world's largest terrestrial carbon banks – and one of the world's largest sources of fresh water. Despite the on-going war on climate change, and despite the well known looming crisis in fresh water supplies and despite massive local protests in both Alberta and the U.S., the stalled but not defeated 1700 mile extension of the Keystone XL Pipeline (owned by TransCanada) would carry still more of Alberta's tar sand oil through the U.S. – all the way through America's heartland into Texas and the Gulf of Mexico. In addition, the new economics of bringing these unconventional fuels to market together with a green light from Washington has allowed yet another Canadian company called U.S. Oil Sands (formerly Earth Energy Resources) to become the first large scale tar sands mine in the United States. This mining operation began June 2013, on leased land in the Tavaputs Plateau of Utah, despite years of local and regional resistance.

What might persuade the Canadian and U.S. governments to give the green light to such seeming insanity? The answer lies in various clauses of the "free" trade agreements. For example, NAFTA's energy proportionality clause requires Canada to provide 2 units of export for every unit produced for domestic consumption. Meanwhile, Chapter 11 of NAFTA allows foreign corporations to sue member governments directly for lost future profits, and both CAFTA and NAFTA have caps on awards for environmental damage – but investor suits are unlimited. In other words, should they be denied the right to open their U.S. mine, the Canadian-owned U.S. Oil Sands has the right to sue the U.S. government for an unlimited amount over loss of future profits, potentially tying up the government in a costly suit for years. At the same time, suits against U.S. Oil Sands for environmental damages are capped.

All is reinforced at the local level by any number of organizations in any number of ways. One such organization is the forty year old American Legislative Exchange Council or ALEC, which works to promote special interest "model" legislation using corporate-funded "scholarships" to send legislators on trips with corporate lobbyists. Thus it was that in October 2012, nine U.S. state legislators attended the "ALEC Academy" which included an industry paid trip to explore the Alberta Tar Sands. Expenses were paid by TransCanada, which is a member of ALEC, along with other groups and corporations with direct financial interest in the Alberta Tar Sands and the proposed Keystone XL pipeline.

Washington's lobbyists employ similar methods, but instead of offering up model legislation to state and local governments as ALEC does, Washington lobbyists seek to modify proposed federal legislation "at the margins" which is where former lobbyist extraordinaire Jack Abramoff says the real power of the Washington lobbyists lies. They get there by providing perks and all expense paid trips to legislators.

Various departments, agencies and activities of the United Nations also play a role in fostering today's special brand of hyper-globalization and providing assistance with globalization's requisite "free" trade agreements. For example, the online organization chart of the UN shows the WTO as a "related organization" and its sustainable development page informs us that "Agenda 21 is a comprehensive plan of action to be taken globally, nationally and locally by organizations of the United Nations System, Governments, and groups in every area in which human impacts the environment." This plan was created in 1992 by an unelected Commission on Sustainable Development. In an article titled *UN's Agenda Targets Your Mayor* one independent journalist provides some insight into how the plan is put into motion at the local level:

> "From June 1 through 5, 2005, the city of San Francisco was the site of an international conference called "World Environment Day." But the agenda of this conference was much bigger than just another hippy dance in the park. This meeting of the global elite had a specific target and an agenda with teeth. The goal was the full implementation of the UN's Agenda 21 policy called Sustainable Development, a ruling principle for top-down control of every aspect of our lives – from food, to health care, to community development, and beyond. This time, the target audience is our nation's mayors. The UN's new tactic, on full display at this conference, is to ignore federal

> and state governments and go straight to the roots of American society. Think globally – act locally. . .
>
> Here's a quick look at a few of the 21 agenda actions called for. Under the topic of energy, action item number one calls for mayors to implement a policy to increase the use of "renewable" energy by 10% within seven years. Renewable energy includes solar and wind power.
>
> Not stated in the UN documents is the fact that in order to meet the goal, a community would have to reserve thousands of acres of land to set up expensive solar panels or even more land for wind mills. Consider that it takes a current 50 megawatt gas-fired generating plant about 2-5 acres of land to produce its power. Yet to create that same amount of power through the use of solar panels would require at least 1,000 acres. Using wind mills to generate 50 megawatts would require over 4,000 acres of land, while chopping up birds and creating a deafening roar. The cost of such "alternative" energy to the community would be vastly prohibitive. Yet, such unworkable ideas are the environmentally-correct orders of the days that the mayors are being urged to follow." (10)

The original comprehensive plan known as Agenda 21 has, as the above article excerpt indicated, evolved over the years. An excerpt from a report titled "Green Grabbing Our Future" summarizes the situation as of 2012:

> "The Rio process itself has been steadily privatized under the weight of 20 years of neoliberal globalization. As the global contradictions between

economy and environment have intensified, nature itself is becoming a source of profit. . . What was once a state-oriented, regulatory framework has morphed into a market-based, corporate initiative. Convinced of the power of the global marketplace to conserve biodiversity, Rio+20's institutions are pushing the commercialization of ecosystem services through market mechanisms like markets for ecosystem services, carbon derivatives, and the 'sustainable' certification of forests, soybeans, palm oil and agrofuels.

The corporate trend to privatize and commercialize ecosystem services and resources in the name of environmental protection is known as "green grabbing" as these schemes can result in local communities losing resource rights. . . It is the favored approach of the big conservation organizations like World Wildlife Fund (WWF), Conservation International (CI) and the International Union for the Conservation of Nature (IUCN), who have thus guaranteed their place at the Rio+20 negotiating table alongside neoliberal governments and powerful multinational business interests.

In contrast, in the UNCSD's largest parallel event, *The People's Summit* social movements fighting for food sovereignty, climate justice, environmental justice, human rights and the rights of indigenous peoples, women, youth and labor are actively protesting official Rio+20 proposals and proposing grassroots alternatives. . .

Almost everyone inside (and outside) Rio+20 agrees that a truly sustainable agriculture can and should increase yields, and help farmers adapt to and mitigate climate change by either reducing emissions or sequestering carbon. . . There is profound disagreement, however, on whether Rio's combination of carbon markets, bio-char and agrofuel plantations and "sustainable intensification" (e.g. micro doses of fertilizers and GMO seeds) will actually deliver social, economic and environmental benefits, or whether they simply open up agricultural resources to financial speculation for corporate profit. . .

The Green Economy concept that determines the content of all submissions [for the Zero draft report] was itself created by a group led by Pavan Sukhdev a former senior banker from Deutsche Bank and head of UNEP's Green Economy Initiative. This is a reflection of a long trend in partnering between the CBD, big environmental organizations and corporate representatives i.e. the World Business Council on Sustainable Development, the International Chamber of Commerce, CI, WWF, IUCN etc.

The purpose of these partnerships is to 'make nature pay for its own conservation' by turning it into a new commodity and by creating new markets where it can be traded through mechanisms like REDD+/Reducing Emissions from Deforestation and Forest Degradation. (This mechanism is a carbon credit market for avoided emissions from deforestation and forest degradation. With REDD+ polluters can get commercially tradable credits for not causing carbon pollution they theoretically would have emitted in the future.)

> The dubious justification for bringing nature to Wall Street—where credits and shares of ecosystem services, biodiversity derivatives, avoided emissions and even wildlife species banking can be chopped up, repackaged and resold along with debt, mortgages, hedge funds and the like—is that the best way to save nature is to sell it. In doing so, we are told, we will grow the economy and this in turn will benefit the poor, thus ending poverty and hunger." (11)

Long before Agenda 21 was ever finalized yet another unelected commission known as Codex Alimentarius, which is responsible for the development of an international code of food standards, was created by the UN's Food and Agricultural Organization (FAO) and the UN's World Health Organization (WHO) in 1962. It's purpose: to "harmonize" world food trade. The proposed rules under Codex seek, in a series of steps, to "regulate everything from farm to fork" throughout the world. Due in part to the undemocratic way in which the Codex Commission has been structured (and not unlike that manner in which Agenda 21 and its green initiatives as well as the free trade agreements have been developed) "voluntary" Codex Rules in practice have nothing to do with protecting or promoting human health and everything to do with lining the pockets of multinational food and chemical corporations.

Relying heavily on Codex Commission recommendations, most if not all free trade agreements contain what is known as the SPS (Sanitary Phytosanitary Measures). Article 3 of SPS states: "To harmonize sanitary and phytosanitary measures on as wide a basis as possible, members SHALL base their food safety measures on international standards, guidelines or recommendations." Thus, as per its own website, *any nation who is a member of the WTO and who may wish to apply stricter food safety measures than those set by Codex may be required to justify these measures scientifically.* The question that immediately comes to mind is whether said justification would be required to rely solely or most heavily on corporate-sponsored science. Moreover, can there be any doubt that – should a member nation wish to apply stricter food safety measures – the move to do so might in all probability subject such a member to "loss of future profits" lawsuits by the likes of Conagra, ADM or other behemoths?

One example of the way these initiatives become embedded within government is this official statement on "harmonization" which comes to us compliments of the Center for Food Safety & Applied Nutrition of the FDA:

> "The harmonization of laws, regulations and standards between and among trading partners requires intense, complex, time-consuming negotiations by CFSAN officials. Harmonization must simultaneously facilitate international trade and promote mutual understanding, while protecting national interests and establish a basis to resolve food issues on sound scientific evidence in an objective atmosphere. Failure to reach a consistent, harmonized set of laws, regulations and standards within the free trade agreements and the World Trade Organization Agreements can result in considerable economic repercussions."

By far, the most draconian of the free trade agreements, known as the TransPacific Partnership or TPP, (along with its marginally less aggressive counterpart known as the Transatlantic Trade and Investment Partnership or TTIP aka TAFTA) is – as of this writing – still under discussion. Lori Wallach, attorney, trade expert and author of *Whose Trade Organization* says that more than 600 official corporate "trade advisors" have access to TPP text drafts, while the U.S. public, the Congress and the press have been denied the same access. The worst part, says Wallach, is that the TPP is about far more than trade. Of the 29 leaked draft chapters, only 5 even remotely deal with traditional trade issues. The scope of the TPP, says Wallach, is more "behind the borders" than any of the previous trade agreements.

While being branded a trade agreement, its functions are closer to what might best be described as "diplomatic legislation" as devised by trade negotiators who think of their constituencies as being comprised of just the big U.S. businesses who are looking to dismantle regulations or buy up companies in other countries. These negotiators are also negotiating changes to all U.S. domestic laws behind our own borders, and this feature in effect represents a kind of slow motion coup d'etat via trade agreements. FOIA requests have been denied on the basis of the TPP agreement being classified as a national security issue. This despite the fact that every other country has access to the drafts, as well as the U.S. position in these drafts.

International tribunals within the TPP would allow individual foreign corporations to be raised to the same level as our whole sovereign nation, and directly sue our government with equal status in order to privately enforce a public treaty. As a result, the 200 cases now pending under other agreements promises to expand exponentially under the TPP since it adds 10,000 corporations to the list that can sue the U.S. Banks of course would be included among the prime beneficiaries since the agreement would allow financial firms to directly attack government polices and demand taxpayer compensation for policies they claim undermine their expected future profits.

In essence, most of the provisions would set rules on non-trade matters that affect our daily lives. These include issues such as food choice, water services, financial regulation, jobs, internet freedom, medical prices and more. Moreover, all – not just some – of our domestic policies and laws would be required to comply with TPP rules.

The Institute for Trade and Agricultural Policy aptly sums up the situation insofar as food systems are concerned:

> "The TPP is not only about lowering tariffs. It has the potential to greatly expand protections for investors over those for consumers and farmers, and severely restrict governments' ability to use public policy to reshape food systems. The fundamental causes of recent protests across the globe over food prices, the rising market power of a handful of global food and agriculture corporations, as well as the dual specters of rising hunger and obesity around the world, point to the need to transform the world's food systems, not to lock the current dysfunctional situation in place." (12)

The Obama administration plans to sign the TPP by the fall of 2013.

The Origins of True Wealth

It was in 1971 – well before the recent spate of free trade agreements – that Charles Walters revisited Jefferson's lifelong philosophical usufruct argument in his book *Unforgiven.* He begins by quoting John Stuart Mill from Mill's scholarly "Principles of Political Economy," which in effect argued that distributions in an institutional world had nothing to do with economics: "Even what a person has produced by his individual toil, unaided by anyone, he cannot keep, unless by the permission of society. . . .*The distribution of wealth, therefore depends on the laws and customs of society.* The rules by which it is determined are what the opinions and feelings of the ruling portion of the community make them, and *are very different in different ages and countries,* and might be still more different, if mankind so chose. . ." [italics mine]

Walters then quotes fellow Raw Materials Economist Vince Rossiter who was commenting on Mill's treatise: "If society did not like the natural results of its activities, it had only to change them. . . But whatever it did, there was *no correct distribution* – at least none that economics had any claim to fathom. There was no appeal to laws to justify how society shared its fruits. *There were only men sharing their wealth as they saw fit. It was a discovery of profound consequences, for it lifted the whole economic debate from the stifling realm of impersonal and inevitable law and brought it back into the arena of ethics and morality.*" (13) [italics mine]

It should be mentioned that Vince Rossiter was a disciple of Carl H. Wilken, who in turn was the subject of Walter's book *Unforgiven.* Similar to Irving Fisher, Wilken's goal was to find the real cause of the Great Depression, and how future depressions might be prevented. He "began his analytical studies with no pre-conceived ideas or economic theories," scientifically analyzing the economic records of the U.S. Government. (14)

Assisted by a team of like-minded Raw Materials Economists, Wilken "discovered a natural law of economics, based upon arithmetic and physics, that had escaped generations of supposedly learned economists. They proved that raw materials income governed national income *unless the latter was expanded by debt.* [Wilken's] data also made it clear that expansion of trade *beyond income* destabilized the internal U.S. Economy . . ." (15) [italics mine]

In an article titled *The Nature of Money*, published in the November 2003-February 2004 "Acres USA Newsletter," Charles Walters fleshes out some of the details discovered by Wilken and his associates. He also points out that all modern political systems – whether they be categorized as fascist, communist, socialist, capitalist or whatever – operate within the same monetary system. The problem that goes unrecognized by almost everyone is that this type of monetary system is always dominated by debt instead of true income, and the eventual outcome will therefore always be the same: wealth being continually extracted from both land resources and small producers and becoming concentrated in fewer and fewer hands. In the process, political power concentrates in those same few hands.

As we have seen, the extraction and wealth redistribution process of our debt reliant monetary system was already in evidence in ancient Mesopotamia, yet at the time of the American Revolution it still dominated the beliefs and actions of the old nobles of Europe and England. Little of this history escaped the founders' notice, and many were vehemently opposed to its replication in the newly created Republic, this because of the damaging effects it had on society. For example, in a book finalized in 1811 but finally published in 1814, John Taylor of Caroline (who was a long time ally of Thomas Jefferson's) explains the wealth redistribution process – and anticipates the consequences – when money is created as debt via a banking system allowed to create the nation's currency:

> "If no new banks should be created after 1808, nor the acquisitions of the old increased, the five millions annually collected by the existing banks, at compound interest, carry from the public to the corporations, in twenty years, above one hundred and eighty-four millions of dollars. Here is already a vast current of money and power running one way; will those check it in whose favor the current sets? Are the receivers, as regulators of power and wealth, of undoubted confidence?"

This is why, like Taylor, Jefferson and their allies, Walters took a strong stand on who is to control the nation's money supply, observing that "surely there is no solution to our economic dilemma without a sound money supply. This is not my [Walter's] observation alone. It is contained in Frederick Soddy's *Wealth, Virtual Wealth and Debt* and it is the backbone of my own work *Unforgiven*."

Finally and again echoing Jefferson's perspective on usufruct, Walters asserts that "the creation of money ought to be in harmony with the creation of raw material, not debt, in the economy." (16) This can only be done in a monetary system in which the legal value of money equals the actual value of a nation's wealth – and NOT the debt and interest it takes to secure them.

A major part of the problem today is that too many of us have been schooled by what William Grieder termed deep lobbying, accompanied by propaganda disguised as education and promulgated by hired gun experts. The results, as explained by Charles Walters, is that:

> "The world understands what it means to go broke, but there seems to be an elusiveness about 'the process' of going broke, just as earning one's way is one of those concepts people misplace from time to time. The main reason officials have no clear cut idea on why a national economy fails to earn its way is because they have no clear cut idea of what 'earning its own way' is in the first place. Because there is so little understanding about earning the income of an economy, it has become a matter of simple expediency to inject borrowed money against profits and savings – relying on the turnover of borrowed money to create prosperity.
>
> Similarly, there is no real understanding nowadays of how unearned income can be spent and still owed for, and accordingly there gets to be a psychology that leads people to believe they can enjoy prosperity indefinitely by mortgaging the future of their children, grandchildren and great grandchildren." (17)

Yet, as we all know, borrowing against our future via a debt reliant monetary system is exactly what we have been doing for over two hundred years, contrary to Article I Section 8 of the U.S. Constitution. This is in large part due to the well organized efforts of the "aristocratic" or "well born" Hamiltonians – whose guiding principle was "that governmental interference with economic laws is desirable when it aids business, but intolerable and unsound when it aims at business regulation or control, or when it assists [small-scale] agriculture or labor."(18) Despite periodic waves of public resistance, this guiding principle has been followed ever since.

The creation of the Federal Reserve System in 1913 was accomplished by the same internationally oriented commercial and political interests that helped sweep Hamilton's plan into being, and contained many of the same elements. This time around the large industrial corporation had, as early as 1903, already become the dominant structural element in the American economy.

What were known as "combinations" or trusts up through the 1880's evolved into more formal legal corporations after the Sherman Anti-Trust Act. These corporations then joined with each other or created new corporations altogether in order to form new "combinations" up and down and across their supply, production and distribution chains as a means by which further control over production and distribution could be achieved. Need it be said that their economic and political clout grew accordingly.

It was in this manner that "the tendency toward what contemporaries called combination or concentration – what historians now know as horizontal integration – was accompanied, beginning on 1900-1901, by an equally impressive integration of extraction, production, and distribution – what historians and economists know as vertical integration, or backward and forward linkages between raw material recovery at one extreme and marketing (advertising, sales, etc) at the other." (19) Over the years this horizontal and vertical integration has only intensified, thanks to the Federal Reserve System. Today's farm and economic justice activists for example have been taking note of the fact that the types of organizational alliances created by Big Ag, Big Oil and Biotech are actually restructuring agribusiness itself, with much of the new money increasingly being devoted to the construction of ever more fully integrated networks, through which production, shipping, processing and distribution are funneled.

Beginning around 1890 and as a means by which to head off Populism (which demanded a dollar with debt-paying power) as well as small entrepreneurial capitalism, a carefully constructed coalition of corporate businessmen and eastern financiers, together with hand picked intellectuals, academics and members of the press began to expound on the blessings of centralized control of production and distribution and the need to create and perfect the corporate instruments of such control. It was through the efforts of this coalition that the Federal Reserve Banking System was eventually created. Through it a new, more highly organized system of corporate capitalism emerged, one in which the savings of society as a whole were to be converted into particular productive instruments, and which was to be managed in accordance with the capabilities and requirements of the large corporations. (20)

This new economic system was in its essence a hybridization of that devised by Hamilton over two centuries ago. Similar to Hamilton's system, the savings of society are accumulated via government and corporate bonds and other securities, then amplified by debt. Thus, when the Fed purchases these bonds or other securities it creates "reserves" which enable the banks to expand the debt/money supply by creating loans based on the money you and I have on deposit – using fractional reserve deposit expansion (and thereby putting our deposits at risk). The big Wall Street banks are first in line for accumulation of "reserves" and it is they, together with the investment community at large, who are thus able to supply the large amounts of capital needed by large corporations for expansion. (21)

The democratization of the stock market helped increase public support for the overall plan, even as it allowed increased accumulation of savings through Wall Street. So deeply embedded into the economy has our democratized stock market become that the nation's pension funds are now stuffed with investment portfolios created and managed by Wall Street, to whom huge procurement and management fees are paid every year. These portfolios are dominated by all the familiar big players, and include Big Pharma, Big Ag, Defense, and Big Energy, with its "Green" Energy subsidiary. Unfortunately for some, many previously "fully funded" pension systems that had been filled to the gills with stock assets in the form of derivatives, especially mortgage backed securities, subsequently became "underfunded" (per Wall Street dictates) when the 2008 housing bubble burst, bringing Wall Street asset prices, many pension plans and the entire economy down with it.

In the three or four decades prior to the creation of the Fed in 1913, big "C" corporate capitalism began to eclipse small "c" entrepreneurial capitalism. This process was in fact made possible by a national income that had long been expanded by debt, helping to "eat up the usufruct" because the legal value of money exceeded the actual value for nearly all of our nation's history. The boom and bust cycles characteristic of debt dominant monetary systems and which likewise have plagued America for all of her history also allowed for the continual transfer of the nation's wealth into fewer and fewer hands. This process has occurred with and without a national bank due to the reluctance of Congress to exercise its duty to coin the nation's money and regulate the value thereof.

We might say that both the Hamiltonian capitalists and the new corporate innovators were what Henry George would have called "the accumulators." These accumulators firmly believed that civilization could not advance – or even exist – without "massed capital." To their way of thinking, civilization itself was at stake if small "c" capitalism and the popularity of "democratic," government-issued currency were allowed to take hold. Similar to the Hamiltonians who sought to maintain the old social and economic order established by Britain and Europe, these new corporate innovators "clearly understood that they were redefining the social context of economic change, and thereby revising relations among social classes. . . [The] men who saw the new corporate economy as a 'natural result' of industrial evolution also saw it as a social product."(22)

Thus the new economy developed through the new Federal Reserve banking system would be dominated by and organized around the new industrial corporations, supporting and expanding horizontal and vertical integration – and effectively marginalizing what William Jennings Bryan referred to as the "broader class of businessmen," which included actual and potential small entrepreneurs, including small farmers. The old Jeffersonian beliefs that large industrial corporations threatened the liberties of the people, together with the political economy of republicanism (which up to that time had been an ideal embraced by the majority of Americans) was neutralized by the promise of secure, good paying jobs offered by the corporate innovators.

For a while the Federal Reserve System seemed to work quite well, especially if one could be persuaded to overlook, or rationalize away, the social costs of the Great Depression and numerous intervening recessions – as well as the public works projects more rightfully described as war and empire building. The Federal Reserve System worked, that is, until the "exponential laws of increase" so familiar to Babylonian mathematicians began to intensify and further concentrate the "vast current of money and power running one way" as John Taylor of Caroline predicted over two hundred years ago. This concentration of power has of course enabled the corporate and financial elite to write history and control media coverage – not to mention business school curricula and our own government – so as to serve their own purposes.

Today, this control of business school curricula means, as economist Michael Hudson clearly explains, that "orthodox academic models rarely acknowledge the problems caused by the exponential growth of debt overhead." Firmly anchored in the belief that economies can simply "grow their way out of debt" like some sort of perpetual motion machine that operates independently of finite land, such models have, by necessity, developed into an even more corrosive post-industrial system that "offers only the virtual (and temporary) wealth of asset bubbles." The resulting "bubble economy" has been described by Hudson as being a two part play for privatization, morphing into what he has most recently dubbed as the "toll booth economy." (23)

Clearly the "one way current of money and power" is raging out of control, with five or six too-big-to-fail Wall Street banks effectively holding our government and our people hostage to a monetary/economic system which threatens to destroy the world. The size and international reach of the U.S. economy and the scope of the power of the too-big-to-fail banks is such that they have the very real ability to bring not just the American economy but the world economy to its knees. The so-called "free" trade agreements are simply part of the legal matrix needed to maintain the "blessings of centralized control over production and distribution" even as they expand the rights of creditors "to subordinate labor and capital to the economy's financial dynamics."

8

LAND FRIENDLY MONETARY POLICY

When one thinks about the most important inventions our modern world has inherited from the ancients, one must conclude that money ranks right up there with agriculture, the wheel and fire. The nation's founders, who were themselves extraordinarily well versed in history, most certainly would have agreed. Their writings and their own experiments with colonial scrip (paper money) clearly tell us that they viewed money as the "lubricant" that allows society to function and progress.

These factors coupled with the founders' universal awareness of the landmark decision of the world famous *Mixed Money Case* of 1604 – giving the state the power to create money – meant that they fully understood that money is a public utility designed for public use. As such, a law – created by the state – is needed to define money. Hence Article I, Section 8 of the U.S. Constitution charges Congress with the "Power To . . . coin money [and] regulate the Value thereof . . ." Article 1, Section 10 *limits* what the states can use as money: "No State shall emit Bills of Credit, make any Thing but gold and silver Coin a Tender in Payments of Debts. . ." (It should be pointed out that no state has ever paid its debts with gold or silver.)

The importance of the *Mixed Money Case* has from the start been almost wholly overshadowed by propaganda emanating from those whose interests the Mixed Money decision, as well as the true colonial experience, does not serve. Thus a summary by Alexander DelMar, one of the nation's greatest monetary historians, appearing in his treatise *The Science of Money* is useful:

> ". . .The Mixt Moneys case [of 1604] decided that Money was a Public Measure, a measure of value, and that, like other measures, it was necessary in the public welfare that its dimensions of volume should be limited, defined and regulated by the State. The whole body of learning left us by the ancient and renascent world was invoked in this celebrated dictum: Aristotle, Paulus, Bodin and Budelius were summoned to its support; the Roman law, the common law and the statutes all upheld it; "the State alone had the right to issue money and to decide of what substances its symbols should be made, whether of gold, silver, brass, or paper. Whatever the State declared to be money, was money."

Every one of the colonies had experimented with colonial scrip, or paper money, created by the respective government as a public utility for public use. In his *Modest Inquiry into the Nature and Necessity of a Paper Currency* published in 1729, Benjamin Franklin discusses the positive effects that "abundant" paper money would have on farm prices, labor wages, manufacturing and trade. He also warns against confusing gold or silver with money asserting that "Silver and gold . . .[are] of no certain permanent value. . .We must distinguish between money as it is bullion, which is merchandise, and as by being coined it is made a currency; for its value as merchandise, and its value as a currency are two different things."

Franklin was not only in keeping with the findings of the Mixed Money Case that money was a public utility created for public use, but he also was echoing the earlier sentiments of John Locke, who had stated in his 1718 *Essay on Money and Bullion*: "Observe well these rules: It is a very common mistake to say that money is a commodity. . . Bullion is valued by its weight. . . money is valued by its stamp."

Along the same lines, Thomas Jefferson would state, in 1786, in response to questions asked by French author Jean Nicholas D'Meunier who was then writing a section on the United States for the *Encyclop die m'thodique*:

> "With respect to bills of credit, wherever state legislatures laid taxes to bring in money enough for that purpose, and paid the bills punctually . . . paper money was in as high estimation as gold and silver."

As Jefferson alludes, there were colonies whose paper money depreciated due – ostensibly at least – to egregious mismanagement, prompting the English Parliament to pass "an Act to regulate and restrain Paper Bills of Credit" in 1751, and again in 1764 when it banned the use of paper bills of credit as legal tender. However, these measures created such a backlash that in 1773 the English Parliament passed "an Act to explain and amend" the 1764 measure. As per the amendment, the colonies would henceforth be able to issue paper money for "the publick Advantage" and "in Justice to those Persons who may have demands upon the publick Treasuries in the said colonies for services performed." This money could be made "a legal Tender to the publick Treasuries" in payment of taxes and other dues. This concession obviously came too late for Britain to escape the Revolutionary War, but it did mean that this particular criticism of British interference with domestic law-making meant that it would not be included in the list of grievances itemized in the Declaration of Independence. (1)

During the Revolutionary War, the Articles of Confederation gave the Continental Congress power to issue U.S. money, but not the power to regulate it as the individual colonies were able to do (when they so chose). At the same time that the Continental Congress issued Continentals, the colonies also were issuing (in an amount commensurate with that of the Continentals) their own disparate currencies. The problem with these currencies, unlike the Continentals, is that they were not always accepted by neighboring colonies, and if they were, their stamped valuations were not uniformly honored. Britain added to this monetary chaos by pumping out about one billion dollars worth of counterfeit Continentals aboard warships off the New England coastline, and advertising their availability on broadsides within the colonies.

These combined factors contributed to a growing lack of faith in all colonial money, which James Madison addressed in his 1779 "Report on Money." In this Report, he correctly asserts that the value of a state (or nation's) money is dependent upon "the credit of the state issuing it," that is to say, the state's willingness to pay its bills in prompt fashion using well-managed currency. In addition, his discussion of the much maligned Continentals accurately assesses the one key reason for their decline in value, which had to do with growing distrust of the public credit caused, as per Madison, by "a combination of enemies employing every artifice to disparage [the currency]."

The problem of monetary confusion within the colonies themselves was compounded by the fact that interest-bearing notes, provided by private parties and nascent banks, as well as by the colonies and the Continental Congress, were circulating (and oftentimes hoarded by those who could afford it in anticipation of future gain), along side of non-interest bearing bills of credit. Madison discusses this problem in his Report by noting that the key difference between a loan-office certificate and a common bill of credit was the interest charged on the loan-office certificate. The use of loan office certificates, Madison wrote, meant that "Instead of paying off the capital to the public creditors, we give them an enormous interest to change the name of the bit of paper which expresses the sum due them; and think it a piece of dexterity in finance, by emitting loan office certificates, to elude the necessity of emitting bills of credit."

Similar to Madison's "combination of enemies employing every artifice to disparage the currency" Jefferson provided the following statement in 1786 for French author Jean Nicholas D'Meunier who was then writing a section on the United States for the *Encyclop die m'thodique*:

> "Those who talk of the bankruptcy of the U. S. are of two descriptions. 1. Strangers who do not understand the nature & history of our paper money. 2. Holders of that paper-money who do not wish that the world should understand it."

Jefferson, like Madison and many of his contemporaries, had long been a serious student of monetary matters. For example, by 1776, he had already drafted a detailed report for Congress on the problem of unifying standards and values of the various currencies used in the colonies. As previously mentioned, the big problem during this time period was that no one knew what any of these individual currencies were worth, creating a situation where one currency would be valued at one amount in one colony and a different amount in another, a situation that was then repeated by the Civil War South. In other words, the "promiscuous circulation" of disparate and sometimes incompatible currencies had by this time become a real problem for colonies as they tried to keep their individual economies functioning.

Jefferson's investigations for this 1776 report led him and many others to conclude that a common United States currency was essential to the future success of the country, as well as that of individual colonies. However, his report was tabled until 1782, at which time Congress asked Robert Morris to "devise a table of rates by which the various currencies could be exchanged at the national treasury." Jefferson balked at Morris' complicated plan, and though uninvited, "threw together" an alternate plan which he published as *Notes On The Establishment Of A Money Unit, And Of Coinage In The United States.* Again uninvited, he registered his alternate plan of 1785 and its revised counterpart about a year later. It is noteworthy that many of his ideas were in fact adopted and remain with us today, including the concept that money is an abstraction and not a "thing." Over the next crucial decades he continued to be an important contributor to the monetary debate, even while he was serving in France between the years of 1784 to 1789.

It is more than a little interesting, and perhaps somewhat curious that in 1785, Madison, Jefferson and Washington's own state of Virginia passed a law preventing the circulation of private bank notes, making it unlawful "for any person to offer in payment a private bank bill or note for money." (2) This however did not deter Hamilton from his plan to circulate bank notes created by his proposed, mostly privately owned Bank of the United States as "money." Echoing the Mixed Money Case, he defended such bank notes as follows:

> "It is immaterial what serves the purpose of money, whether paper or gold and silver; that the effect of both upon industry is the same; and that the intrinsic wealth of a nation is to be measured, not by the abundance of the precious metals contained in it, but by the quantity of the productions of its labor and industry."

Hamilton may have understood the function and purpose of money but he seems never to have considered the usufructory limits of the land. Similarly, his grandiose ambitions for the U.S. may have prompted him to overlook the deleterious effects that the "exponential laws of increase" were even then foisting upon England and her people. Few however were fooled by Hamilton's initiative to establish bank notes as the nation's currency. As then Senator William Maclay of Pennsylvania, who has sometimes been referred to as the first Jeffersonian, wrote in his journal in 1790: "Bank bills are promissory notes, and, of course, not money. I see no objection in this quarter." In Maclay's mind, as in the minds of the Jeffersonians, no one should seriously argue that bank bills were the same as money.

Unlike Hamilton, we can see from Jefferson's own words that he, perhaps more than any others of his generation, clearly understood that there were indeed usufructory limits on the land. Hence his lifelong belief that no generation had a moral right to bind another with financial claims that exceeded the present generation's ability to pay. Throughout his life he rightly made the point, over and over again and in numerous ways that bank notes are not money, but rather only "nominal money." Our reliance on such "nominal money," whose legal value must by its very nature exceed the actual value of the nation's capital wealth, means that the unpaid debts of one generation will force succeeding generations to draw unnecessarily from nature's capital reserves – using both human labor and the produce of the earth – simply in order to pay the interest on the debt foisted on them by the previous generation.

So it was that in 1801, Jefferson asked his Treasury Secretary Albert Gallatin:

> "In order to be able to meet a general combination of the banks against us in a critical emergency, could we not make a beginning towards an independent use of our own money, towards holding our own bank in all the deposits where it is received, and letting the treasurer give his draft or note for payment at any particular place, which, in a well-conducted government, ought to have as much credit as any private draft or bank note or bill, and would give us the same facilities which we derive from the banks?"

So it was that in 1813 Jefferson was still urging monetary reform for the same reasons of protecting the usufruct. This is, for example, evidenced in one of a series of similar letters that he would write to his son-in-law John Eppes, who was then serving as Chairman of the Committee of Ways and Means of the House of Representatives: "The truth is, that capital may be produced by industry, and accumulated by economy; but jugglers only will propose to create it by legerdemain tricks with paper." And in another letter of the same year he offered Eppes the following advice:

> "The question will be asked and ought to be looked at, what is to be the resource if loans cannot be obtained? There is but one, "Carthago delenda est."[i.e., the resilient money power must die.] Bank paper must be suppressed, and the circulating medium must be restored to the nation to whom it belongs. It is the only fund on which they [the public] can rely for loans; it is the only resource which can never fail them, and it is an abundant one for every necessary purpose. Treasury bills, bottomed on taxes, bearing or not bearing interest, as may be found necessary, thrown into circulation will take the place of so much gold and silver, which last, when crowded, will find an efflux into other countries, and thus keep the quantum of medium at its salutary level. Let banks continue if they please, but let them discount for cash alone or for treasury notes."

Similarly, in an 1815 letter to William Crawford, Jefferson wrote: "Treasury notes of small as well as high denomination, bottomed on a tax which would redeem them in ten years, would place at our disposal the whole circulating medium of the United States." And then in an 1816 letter written to John Taylor of Caroline, Jefferson says "I sincerely believe, with you, that banking establishments are more dangerous than standing armies; and that the principle of spending money to be paid by posterity, under the name of funding, is but swindling futurity on a large scale."

The late T. Cushing Daniel, one of America's leading monetary scholars, would have wholeheartedly agreed with the founders concerning the function and nature of money. In books such as *Real Money versus False Money – Bank Credits* and *The High Cost of Living – Cause and Remedy* as well as a September 25, 1913 statement prepared for the Senate Banking and Currency Committee, Mr. Daniel thoroughly disproves the mistaken notion that "intrinsic value" is an essential characteristic of money and further exposes the manner in which this notion has always been a "transparent fallacy" generally perpetrated by those advocating gold as the basis for money. Declaring correctly that it was the dollar which put the value in gold and not the other way around, (or that it was the stamp that gave the gold dollar its value when it was to be used as money), Mr. Daniel further cautioned, as Jefferson himself might have done, that *"it should be borne in mind that the value of the American dollar does not depend upon bankers or gold, but upon the National wealth of the United States created by the people."*

His book *The High Cost of Living, Cause and Remedy,* published in 1912, was dedicated to the "Honest Workers and Real Producers of the World." The farmer played a central role in his argument. He began his book by declaring – correctly – on the title page, right above the book title, that "The biggest question before the world today is money. Other questions are important chiefly as they bear on this overshadowing problem." In the first two chapters he provides the following yet-to-be-heeded admonishment:

> "The people must save themselves. Their representatives in Congress have been 'weighed in the balance and found wanting.'
>
> Let us [therefore] get away from theories of governments and realize once for all that the earth supports us and whatever embarrasses and retards its productiveness, which is the beginning and the very

> foundation of all wealth, is hurtful to the human family and, if continued, destructive of any form of government.
>
> [The money question] is the fundamental great economic question now before the world . . . other issues are all trivial in comparison. It makes no difference where a man lives; if in a great city, far away from the farm, he yet bears a definite relation to a certain piece of land that feeds him and supports him . . . This being a self-evident fact, that no sane man will deny, why go into all the wild speculative theories of government upon which so-called Statesmen and Politicians ride into office and live upon the taxes of the people?"

From the earliest days of the new Republic, the Jeffersonians butted heads with the money power, and lost. One especially poignant letter written by Thomas Jefferson, (who was reflecting the views of the majority of his contemporaries) to George Washington on September 9, 1792 reveals how the people's representatives in Congress had even then been "weighed in the balance and found wanting:"

> "[Hamilton's] system flowed from principles adverse to liberty, & was calculated to undermine and demolish the republic, by creating an influence of his department over the members of the legislature.
>
> I saw this influence actually produced, & it's first fruits to be the establishment of the great outlines of his project by the votes of the very persons who, having swallowed his bait were laying themselves out to profit by his plans: & that had these persons withdrawn, as those interested in a question ever should, the vote of the disinterested majority was clearly the reverse of

> what they made it. These were no longer the votes then of the representatives of the people, but of deserters from the rights & interests of the people: & it was impossible to consider their decisions, which had nothing in view but to enrich themselves. . ."

Later, after a long fight against a National Bank as the means by which to create the nation's currency and Madison's own plea to Congress to direct the Treasury to create the nation's money, Jefferson writes in an 1817 letter to Josephus B. Stuart:

> "The bank mania is one of the most threatening of these imitations. It is raising up a moneyed aristocracy in our country which has already set the government at defiance, and although forced at length to yield a little on this first essay of their strength, their principles are unyielded and unyielding. These have taken deep root in the hearts of that class from which our legislators are drawn, and the sop to Cerberus from fable has become history. Their principles lay hold of the good, their pelf of the bad, and thus those whom the Constitution had placed as guards to its portals, are sophisticated or suborned from their duties."

Modern monetary reformists face yet another problem today because our daily experience with "nominal money" proves to us that this substitute, "nominal" money, in the form of bank notes, works quite well in terms of facilitating the payment of our bills and making purchases, or in other words providing the lubricant that makes society run. We thus lose sight of the all-important fact that all debt-dominant monetary systems – including our current one – are subject to the exponential laws of increase, thereby forcing us to "eat up the usufruct" of the land, by substituting paper claims on wealth for real wealth.

As we have discussed, mankind has struggled with the social and moral evils of debt dominant financial systems for millennia, resulting in the insidious but certain destruction of our usufruct, disadvantaging countless billions around the globe in the process. Thus, as the writings of Charles Walters, Thomas Jefferson and other independent thinkers reveal and as Stephen Zarlenga of the American Monetary Institute, and author of the ground-breaking book *The Lost Science of Money*, so poignantly and accurately summarizes:

> "The money system is society's greatest dispenser of justice or injustice. A good one functions fairly, helping create values for life. A bad, unjust one obstructs the creation of values; gives special privileges to some and disadvantage to others causing unfair concentrations of wealth and power; leading to social strife and eventually warfare and a thousand unforeseen bad consequences – physical and spiritual." (3)

Circumstances today are such that we must, once and for all, face the fact that we have for far too long been manufacturing reserves out of nothing as a means by which to fictitiously expand nature's base capital reserves, rather than keep them whole and intact. We are fast approaching the point of no return in terms of our ability to "recapitalize." When we delude ourselves into thinking that we, as a society, can create wealth through debt, thus robbing from nature and our own future, we set ourselves up for certain catastrophe.

Clearly we are at a crucial crossroads in human history, but this time fate may be on our side, provided we are willing to seize the opportunity. Certainly the on-going global fallout from the 2008 banking crisis presents us with a rare opportunity for moving forward with much-needed monetary reform. The task has been made easier to accomplish than ever before by a number of fortuitous developments.

One such development is HR2990, introduced September 21, 2011 by former Congressman Dennis Kucinich and Congressman John Conyers. (4) This Bill, written by Congressman Kucinich, contains all the monetary provisions of the 2009 draft of the American Monetary Act which itself represented a culmination of efforts and contributions from Charles Walters and numerous other seasoned monetary scholars, researchers and reformers, under the auspices of the American Monetary Institute.

The American Monetary Act, and in turn the NEED Act, draws from and improves upon the Chicago Plan advanced by Irving Fisher, Henry Simons and other leading economists of the 1930s. It also has a solid anchor in history, based as it is "on Aristotelian monetary concepts in existence since at least the 4th century BC and employed successfully in a variety of monetary systems since then, ranging from democratic Athens to republican Rome. It is not merely a theory – its main elements have a long history of successful implementation in major societies around the world, including the American Colonies and the United States. These concepts enabled us to first establish the U.S. and then to maintain it as one nation." (5)

In 2011, Dr. Kaori Yamaguchi, who holds a PhD in Economic Theory from the University of California at Berkley and a masters degree in mathematical economics from Kobe University, presented his *Workings of a Public Money System of Open Macroeconomies – Modeling the American Monetary Act* to the 29th International Conference of the System Dynamics Society in Washington D.C., and again at a US Congressional Briefing sponsored by Congressman Kucinich, also in Washington D.C., and also at the 7th Annual American Monetary Reform Conference in Chicago. His abstract reads as follows:

> "Being intensified by the recent financial crisis in 2008, *debt crises seem to be looming ahead among many OECD countries due to the runaway accumulation of*

> *government debts. This paper first explores them as a systemic failure of the current debt money system. Secondly, with an introduction of open macroeconomies, it examines how the current system can cope with the liquidation of government debt, and obtains that the liquidation of debts triggers recessions, unemployment and foreign economic recessions contagiously. Thirdly, it explores the workings of a public money system proposed by the American Monetary Act and finds that the liquidation under this alternative system can be put into effect without causing recessions, unemployment and inflation as well as foreign recessions.* Finally, public money policies that incorporate balancing feedback loops such as anti-recession and anti-inflation are introduced for curbing GDP gap and inflation. They are posed to be simpler and more effective than the complicated Keynesian policies. . ." (6) [italics mine]

The very next year, in August of 2012, IMF economists Michael Kumhof and Jaromir Benes came out with a report they called *The Chicago Plan Revisited.* Having had the advantage of the mathematical tools necessary to crunch the numbers that Simons and Fisher, as authors of the Chicago Plan, lacked, it is no exaggeration to say that this paper – together with Professor Yamaguchi's – provides impressive substantiation for the workability, not to mention the wisdom, of the NEED Act/ American Monetary Reform Act. The Conclusion of the Kumhof/Benes paper reads as follows:

> "This paper revisits the Chicago Plan, a proposal for fundamental monetary reform that was put forward by many leading U.S. economists at the height of the Great Depression. Fisher (1936), in his brilliant summary of the Chicago Plan, claimed that it had four major advantages, ranging from greater macroeconomic stability to much lower debt levels

throughout the economy. In this paper we are able to rigorously evaluate his claims, by applying the recommendations of the Chicago Plan to a state-of-the-art monetary DSGE model that contains a fully microfounded and carefully calibrated model of the current U.S. financial system. The critical feature of this model is that the economy's money supply is created by banks, through debt, rather than being created debt-free by the government.

Our analytical and simulation results fully validate Fisher's (1936) claims. The Chicago Plan could significantly reduce business cycle volatility caused by rapid changes in banks' attitudes towards credit risk, it would eliminate bank runs, and it would lead to an instantaneous and large reduction in the levels of both government and private debt. *It would accomplish the latter by making government-issued money, which represents equity in the commonwealth rather than debt, the central liquid asset of the economy, while banks concentrate on their strength, the extension of credit to investment projects that require monitoring and risk management expertise.*

We find that the advantages of the Chicago Plan go even beyond those claimed by Fisher. One additional advantage is large steady state output gains due to the removal or reduction of multiple distortions, including interest rate risk spreads, distortionary taxes, and costly monitoring of macroeconomically unnecessary credit risks. Another advantage is the ability to drive steady state inflation to zero in an environment where liquidity traps do not exist, and where monetarism becomes feasible and desirable because the government does in fact control broad

> monetary aggregates. *This ability to generate and live with zero steady state inflation is an important result, because it answers the somewhat confused claim of opponents of an exclusive government monopoly on money issuance, namely that such a monetary system would be highly inflationary. There is nothing in our theoretical framework to support this claim. And as discussed in Section II, there is very little in the monetary history of ancient societies and Western nations to support it either."* (7) [italics mine]

It was the economic and social wreckage of the Great Depression which prompted the best economic minds in the country to come together to devise what became known as the Chicago Plan, garnering at the time nationwide support from the economics profession. In breaking away from conventional economic thinking the Chicago Plan truly represented a giant step forward in the development of proper monetary reform. As previously mentioned, it also was the precursor to and provided the model for the American Monetary Act, and by extension the Need Act. In and of itself however, it did not go quite far enough.

For one thing, as Dr. Joseph Huber points out in a series of monetary reform critiques, changes (such as those suggested by the American Monetary Act and the Need Act) in the procedures by which money is created and issued along with changes in accounting procedures and management of payment transactions will structurally separate the function of money creation from the functions of lending and finance, something the Chicago Plan did not do. (8) These changes alter the essence and nature of money and actually requires the government-authorized body to spend money into circulation by way of what we might call duly authorized and achievable "public projects" for the benefit and general welfare of the people. Thus as Dr. Huber explains:

> ". . . [Government-issued] debt-free money is embedded in a context of economic obligations. This does not involve a banking debt but a social duty as expressed in modern principles or values such as work, performance, achievement and merit. *Without human effort, labour, technical efficacy and the regenerative forces of nature there is no economic product to sell and buy*, and no purposes in which to invest and build up capital. Money would have no function and be worthless. *Debt-free* [government-issued] *money may not be a promise to repay* [as is the case with Federal Reserve bank notes], *but it is a promise to be productive, and a promise to keep control of the quantity of money so that there is neither too much nor too little money around in correspondence with actual levels of productivity."* (9)

Similarly, in a 1992 book titled *Rights Vs Privileges*, Georgist Robert DeFremery offers "hard" money enthusiasts the following brief discussion on what it takes to maintain a "hard" money standard:

> "There are people who look with distrust upon 'printing press' or 'fiat' money. But they overlook one of the basic facts about money. It is true that we need a 'hard' money. But we should not make the mistake of associating 'hardness' with convertibility into gold. The essence of a hard money is not determined by the material of which it is composed – or the material into which it is convertible. The essence of a hard money is that its supply is fairly stable and there are precise limits to it... And a purely paper or 'fiat' money can be a hard money if we set precise limits to its supply, or it can be a soft money if we set no precise limits to its supply."

As mentioned, the American Monetary Act, like the NEED Act, agrees with the main features of the Chicago Plan but builds upon it. Thus, and in keeping with Professor Huber's discussions above, "the Act proposes that infrastructure expenditures, including education and health and farming parity be used as mechanisms to get newly created money spent into circulation to promote the general welfare." In this way, "money will be created by the government as money, not as private interest-bearing debt . . . and will be monitored to be neither inflationary or deflationary." In addition, and strikingly reminiscent of Jefferson's democratic republicanism, "the Act introduces considerations of fairness, sustainability, sound environmental practice and social cohesion as values in monetary decision making. In other words moral considerations are explicitly considered."(10)

Stephen Zarlenga, director of the American Monetary Institute and author of the game-changing *Lost Science of Money,* provides a very useful summary of the three essential elements of proper monetary reform, and the very crucial need to enact them together as a whole:

> "Monetary reform is achieved with three elements which must be enacted together for it to work. Any one or any two of them alone won't do it, but would further harm the reform process. . . .
>
> *First,* incorporate the Federal Reserve System into the U.S. Treasury where all new money would be created by government as money, not interest-bearing debt; and be spent into circulation to promote the general welfare. *The monetary system would be monitored to be neither inflationary nor deflationary.* [italics mine: see Part 2 of *The Two Faces of Money* by Geraldine Perry and Ken Fousek for a discussion of how this monitoring could take place. Also, in the "Action Corner" of The Two Faces of Money website, under

> "Sue's Page," look at the HR2990 Transition Arrangement flow chart]
>
> *Second,* halt the bank's privilege to create money by ending the fractional reserve system in a gentle and elegant way. All the past monetized private credit would be converted into U.S. government money. Banks would then act as intermediaries accepting savings deposits and loaning them out to borrowers. They would do what people think they do now. *This Act nationalizes the money system, not the banking system.* Banking is not a proper function of government, but providing the nation's money supply is a government prerogative!
>
> *Third,* spend new money into circulation . . . starting with the $2.2 trillion that the Civil Engineers estimate is needed for infrastructure repair; creating good jobs across our nation, re-invigorating local economies and re-funding local government at all levels." (11)

The Need Act, as the legislative configuration of the American Monetary Act, is a truly remarkable document. Brilliant in its design, lucidness and brevity, it would surely have commanded the respect, admiration and appreciation of the Jeffersonians, and not a few of the Hamiltonians as well. Section 501 under Title V provides for direct funding, on a per capita basis, of the nation's infrastructure in order to "modernize, improve, and upgrade the physical economy of the United States in such areas as transportation, agriculture, water usage and availability, sewage systems, medical care, education, and other infrastructure systems."

These projects, being national in scope, would be developed specifically so as to "promote the general welfare" – beginning with repair of the nation's roads and related infrastructure as recommended by the American Society of Civil Engineers. This "public project" in particular would create non-exportable, good paying *private sector* jobs at the local level, all while helping to make the entire economy work more efficiently.

As new money for specific public projects is thus spent into the economy and debt gets paid down, employers and employees would begin to function less as adversaries in a world of monetary scarcity and more like partners in productive work. As a result, collaboration and cooperation would increasingly displace the obscene levels of cut-throat competition and corruption we now see all around us.

Collaboration and cooperation would be further encouraged by the fact that "new money" would no longer be required to first go through the discriminatory filters of the banking system or Congressional earmarks or pork barrel projects before entering the economy. With pork and earmarks out of the picture, politicians and voters alike would feel far less compelled to act solely out of self-interest and self-preservation. Meanwhile, the benefits accruing to local economies would be more swiftly, equitably and widely dispersed throughout the nation, without the exponentially growing burden of debt.

As previously discussed, the exponential laws of increase governing debt pinpoints a major flaw in our current monetary system. One of several ways the NEED Act addresses this is by prohibiting the Treasury Secretary along with federal agencies, departments and other instrumentalities from borrowing from any source other than the U.S. Treasury. In essence, agencies would no longer be able to expand their activities through debt, but instead required to operate on the basis of "full legal tender" obtained through government revenues and as authorized by the people. (Congress of course retains its Constitutional right to borrow money on the full faith and credit of the United States.)

To ensure that the money supply be monitored properly, an *independent, autonomous* Monetary Authority would be created within the Department of the Treasury and under general oversight of the Secretary of the Treasury. This new Authority would "pursue a monetary policy based on the governing principle that the supply of money in circulation should not become inflationary nor deflationary in and of itself, but will be sufficient to allow goods and services to move freely in trade in a balanced manner." Because the new monetary system is no longer biased toward growth, the obscene levels of waste we see all around us will decline dramatically, and Americans would be provided with a dollar of stable value, one that is able to retain its purchasing power over time. This is a huge win-win for both employers and employees, not to mention governmental entities of every stripe and at every level.

Another key component of the legislation is that the Federal Reserve System and all fractional reserve lending would be abolished, with the government purchasing "all net assets of the Federal Reserve System, including the Federal Reserve banks," but leaving in place the all-important check clearing and payment system. Dissolving the present system and ending fractional reserve lending means that privately held corporations would no longer be able to profit from the creation of our national currency. That said, and in keeping with the rule of law, "any reserves of any member bank that is held by any Federal reserve bank shall be returned to the member bank in the form of United States Money."

To replace the current debt/money creation system, a new Bureau of the Federal Reserve would be established within the Treasury Department, for two basic purposes having to do with getting U.S. currency into circulation. First would be to administer "the origination and entry into circulation of United States Money, subject to the limitations established by the Monetary Authority." Second would be to administer "the lending of United States Money to authorized depository institutions." This second provision essentially replaces the Federal Reserve Banking System with the U.S. Government as the "lender of last resort" – thereby ensuring that money creation is solely a function of the U.S. Government and that all fractional reserve lending is ended.

All new bank loans would be made from "100% reserves" of real U.S. money – money that actually is in the bank as savings. This means that irrationally exuberant speculation will be greatly reduced since borrowing is restricted to existing savings, rather than being leveraged through savings, as it most specifically and deliberately has been through the Federal Reserve System.

In addition, interest rates are capped at 8% so no more predatory loans will be allowed – nor will they be needed once 100% US money begins to circulate in the economy in amounts sufficient enough to pay debt down to payable levels and good paying, private sector jobs become more widely available. Compound interest is ended as well, with the rule that total interest may never exceed the principle except on mortgages.

In point of fact, as new U.S. money, itself unencumbered by debt, enters the economy for productive purposes, debt everywhere will be paid down to a much more manageable level, and in the case of the federal debt, eventually eradicated altogether – which of course in turn means that federal taxation levels would decline dramatically. Additionally, the fiscal crisis now permeating federal, state and local governmental entities will come to an end. Social and political unrest, not to mention land abuse, stemming from the extractive nature of the current monetary system will cease.

No longer burdened by unsupportable debt, the economy at every level will transform from one dominated and consumed by debt into one of relative monetary abundance even as the obscene levels of waste and subsequent periods of "creative destruction" created by boom and bust cycles are eliminated. All this and more without raising taxes and without additional borrowing by the government!

The new and widespread availability of local, highly productive, living wage jobs will put an end to Jefferson's fear that our citizens, like those of old England and France, would be reduced through debt and taxes "to such wretchedness, as that laboring sixteen of the twenty-four hours, they are still unable to afford themselves bread, or barely to earn as much oatmeal or potatoes as will keep soul and body together." Needing just one living wage job instead of two, three or even four "as will keep body and soul together," more of our citizens would now have the time and energy needed to more fully participate in local as well as national decision-making processes – this in addition to their own on-going self-education activities. In this and other ways our citizens would be more willing, likely and able to more fully exercise, in democratic fashion, their sovereign rights as citizens, thus realizing the vision of the Jeffersonians and a few of the "aristocratic" Hamiltonians as well.

And since more citizens would have decent paying jobs, more of them would be not just willing – but actually able and more likely – to purchase high quality, nutrient dense, unsubsidized food from responsible local producers whose guiding principle is to rebuild the natural structure of the soil while simultaneously increasing the nutrient value of the food they produce and reducing damage to the environment and water systems, helping sequester CO2 in the soil where it needs to be.

It perhaps goes without saying that the prices these farmers charge must be commensurate with an "income net of soil depletion and cost of production." The fact is that these kinds of responsible, caring farmers do already exist and more of them come on board every year despite enormous odds, but at present the consumer base for these farmers is limited to the more financially blessed among us. This is an unhappy circumstance that will change with proper monetary reform.

Developing in tandem with consumer support of carbon sequestering, regenerative agriculture, the new availability of good paying jobs would also allow more of our citizens to support local food artisans and other small businesses who hitherto could not compete with the artificially low prices created through subsidies and the cheap labor accessed via international "free" trade. Local food artisans, small businesses, craftsmen and trades of all sorts, including locally owned, fully independent media outlets would be able to thrive in this new economic environment. A thriving local economy would also help reduce and perhaps end altogether the present tendency of public officials to coddle behemoth agribusiness operations by overlooking untoward behavior and giving in to unreasonable demands.

Moreover, the consumer-driven choice toward value-added food would provide a self-reinforcing mechanism for responsible small farmers to work with their customers to tailor production and distribution according to more localized needs. Bilateral trade agreements would be used to fill in gaps for particular foodstuff of the two nations in question, and an appropriate, if more limited and continually revolving reserve of high quality, nutrient dense food stocks would be kept in place at the local level for those unforeseen times when weather or other factors curtail normal production. An added side benefit to the move away from plantation-style commodity crops and toward value-added foods is that radical fluctuations in the local money flows that once occurred during harvest season and which were a reason given by the corporate innovators of the late 1800''s to control production, will smooth themselves out.

Inasmuch as money is power, it naturally follows that as debt gets paid down and money becomes more widely distributed, the so-called silent majority will become more empowered politically as well as economically. As the power balance shifts back to the people, more and more of our communities will be able to more effectively participate in, and influence, the public debate even as they seek to refashion those regulatory and administrative laws which grant legal clout to corporations while disadvantaging nature and local communities. Within this new economic environment, there will also be far less need for the seemingly endless parade of new state and federal laws, statutes and regulations which are now deemed necessary to keep the destructive social, economic and environmental effects of the current monetary system at bay.

In this whole process our newly empowered citizens would – with a growing new awareness – join their activist allies in seeking fairer, more economically and environmentally sound trade laws, not to mention a less pecuniary approach to eminent domain decisions and a better adherence to local environmental standards. These newly empowered citizens will, in other words, be better able to join their activist allies in support of political and other leaders who show a willingness and capacity to challenge establishment thinking, perhaps even opening public campaign financing to public debate. Local, *bottom-up* control in other words would begin to replace top-down control, particularly that kind of control coming from unelected policy makers.

With renewed public support, universities and public schools in general will become more independent and less willing to kowtow to the demands of corporate grants and other forms and sources of influence peddling. And as federal, state and local governments themselves become more economically independent, they too will become less willing to provide tax benefits, assist with corporate land grabs via eminent domain – or provide countless other incentives and legal protections to polluters, war propagandists and other land destroyers merely to secure the *promise* of bringing much-needed jobs and revenue into their jurisdiction. The "ears" and minds of educators, government and other officials will begin to bend toward the people they were elected to serve.

More balanced budgets and a newly empowered citizenry also means that the relatively small amount of money needed to bring already proven "new energy" technologies into the public sphere will become available throughout the economy. Once these technologies become widely established there will be no need for wars for oil, gas and similar resources; no need to destroy the boreal forests in order to mine tar sands oil; no need to raise the rainforests in order to supply biofuels or create carbon sinks; no need for hydropower dams which destroy river systems and invite earthquakes; and no more pollution associated with today's conventional and alternative energy choices.

Like regenerative agriculture, the economic, social and environmental benefits of new energy technologies are almost beyond our ability to imagine – but they are real nonetheless. The truth is however that it is highly uncertain, perhaps even unlikely, that they will ever truly come into their own without proper monetary reform. This is unfortunate to the extreme because we are so close, yet so far away from the full embrace of an energy source which, like regenerative agriculture, holds the fullest possibility of healing mankind and the planet on which we live.

All of the afore-mentioned, sorely needed political, economic, environmental and social benefits to local communities – and more – would be further augmented by two other important provisions of the Need Act. One of these provisions, Section 504 of Title V, provides annual monetary grants to the states, on a per capita basis, equal to 25 percent of the money created under Title V in the previous year. Thus the states will be able to determine for themselves where 25% of each years' newly created money will go. At the same time, Section 510 of Title V provides for "interest free lending to State and local governmental entities, including school boards and emergency fire services for infrastructure improvements under their control and within their jurisdictions, based on per capita amounts and other criteria to assure equity as determined by the Monetary Authority."

This last provision alone could save countless billions in interest charges and loan fees now being paid by these various governmental entities. And, insofar as money is power, both provisions would immeasurably help return political power to the people, where it belongs.

The possibilities for achievable, duly authorized public projects made possible by "sovereign" money are myriad, and very much integral to the operations of the new monetary system, as Dr. Huber, Stephen Zarlenga and others explain and as new computerized models demonstrate. Some of these projects are suggested in the Need Act itself. A free, high quality public educational system available to all regardless of income, reminiscent of Jefferson's dream, is one of these public projects.

Another of these projects is a universal health care system, available to all, regardless of income or job status. Following the outline of Dennis Kucinich's "single payer" plan, which was drawn up by four thousand Physicians for National Health Care, this system would be designed for people – not profits (which are an unavoidable priority in the current monetary system). As such it would allow individual freedom of choice of treatments and health provider, including chiropractic and other "holistic" providers – and it would include access to natural remedies and supplements. With this plan and as studies noted by the Physicians for National Health Care showed, health care costs would decrease dramatically.

Need it be said that health care costs would decrease not just through reduction of administrative overhead, profit-taking, and similar factors noted in the studies. Costs would also decrease as the population begins to reap the myriad of health benefits derived from additive-free, nutrient dense, sustainably grown food and a less stressful lifestyle made possible by the ability to prosper on just one living wage job, instead of the two – or more – needed now just to "keep body and soul together."

To help put consumers and their selected practitioners back in control of their health care choices and reduce "top-down control" or micromanagement of the health care system, perhaps a blend of some sort of nation-wide health insurance program limited to serious injury or illness coupled with individual health savings accounts could be established. The major goals would be to ensure that health consumers would have direct access to payments for their selected health care choices thereby establishing more of a true partnership with their selected practitioners.

The major obstacles to all we have discussed so far are without doubt formidable, as they long have been. First, the well established one way current of money and power – as John Taylor of Caroline noted two hundred years ago – will require major shifts in awareness within the electorate in order to effectively reverse the "one way current." This is a daunting task to be sure, because all too often neither those whom the current system disadvantages nor those whom it favors are sufficiently aware of the true root of the problem – which is as this book has discussed, the current debt reliant monetary system. Of course, it goes without saying that those whom the system favors will offer the most resistance. This is precisely why, as former IMF economist Simon Johnson and his co-author James Kwak mention in their book *13 Bankers: The Wall Street Takeover and the Next Financial Meltdown: "We've been here before. The confrontation between financial power and democratically elected government is as old as the American Republic."*

Another related obstacle worth mentioning is that the money power has long had the means and the motive to confuse, confound, and cajole with potential for riches the body electorate with regard to the true nature and function of money. Thus it was that Thomas Jefferson wrote in 1786 that *"Those who talk of the bankruptcy of the U. S. are of two descriptions. 1. Strangers who do not understand the nature & history of our paper money. 2. Holders of that paper-money who do not wish that the world should understand it."*

Certainly it is true that gamblers know that the particular allure of a trifecta bet is that it yields very high returns. Arguably, it is these high returns which have for millennia prompted mankind as a whole to place trifecta bets, gambling that economically driven wars of aggression will take first place, land use second place and monetary policy a distant third. Today a new trifecta has emerged, one which exchanges wars of aggression for a more subtle economic war known as climate change, which will only continue to lead us to all the wrong solutions. In both cases, gamblers and non-gamblers alike have failed to recognize how the current monetary system, which has dominated mankind throughout most of history, not only "eats up the usufruct" but is the very thing now driving unnecessary wars and unwise land use policies, which in turn play a major role in climate.

We have seen how the extractive nature of the current monetary system – which produces more claims on wealth than there is actual wealth available – forces us to destroy soil carbon and humus in attempts to fulfill the contractual obligations of those claims. We have seen how depleted soils actually *require* CO2 in order to produce healthy plants and help nature continuously recycle carbon back into the soil, rebuilding nature's base reserves in the process. We have seen how land use policies, responding as they must to the demands of the legal rather than actual value of money, have over millennia resulted in climate change. We have seen how the current monetary system forces us to pursue legal and economic policies which encourage us to wage a misguided war on climate change that in no way heeds nature's own economic system and the manner in which the land – properly treated – can rebalance greenhouse gases while rebuilding precious, carbon rich humus. We have, in other words, seen how our long-standing trifecta bets have failed us miserably.

Money, as we have shown, and as the nation's founders well understood, is a public utility designed for public use. No one today can deny that money control's society's direction, yet too few recognize that this is precisely why monetary reform must come before all other reform. It bears repeating then that monetary reform is, beyond a shadow of a doubt, the most important question of our time. All other questions are important only insofar as they bear on this long standing issue.

Thus, as John Stuart Mill observed long ago, if society does not like the natural results of its activities, all it has to do is change them. Once we fully understand and embrace the idea that money – like land use policy – is a legal construct set into law by permission of society and once we fully understand and embrace what it will take to install "land friendly" monetary reform, we can at long last put an end to this senseless gambling, allowing humankind and nature to reap the truly extraordinary benefits that have heretofore eluded them.

Thomas Jefferson and his many allies well understood that Hamilton's financial plan, and especially his currency system, would almost irreversibly set the course and direction of the new nation away from the democratic republic it was intended to be and in it its place foster the growth of a new aristocracy populated by rent-seekers, speculators and gamblers of every stripe. Just as the early Jeffersonians feared, from Hamilton's time forward, the far better organized – and motivated – money power, or new aristocracy, has repeatedly won out over unorganized democracy.

Countless times over these two hundred plus years the greed, blatant corruption and cruel oppression of the money power has forced the money question upon the people, with little real success. The difference today is that earth itself quite literally hangs in the balance. The simple, unavoidable truth is, and long has been, that much-needed and long overdue social, land use, environmental and economic change simply can't happen in any realistic fashion without proper monetary reform.

As we all know, or should know, it can no longer be seriously argued that the need for monetary reform is unnecessary. Nor *should* it be argued that legislation such as the Need Act is "ahead of its time." The historical evidence supporting government-issued, land friendly money is now given further credence by reputable modern economic tools and for our convenience it is now presented in concrete form by a stunningly well designed piece of model legislation.

The gambling must end. We have the knowledge, we have the tools – and most urgently we have the need. It can be done and it must be done. There is no more waiting.

POSTSCRIPT

The following is taken from a series of long conversations that took place in 2009 and 2010, just as I was beginning this book, with someone I count as being among my most valued of friends and a truly wonderful human being – who also happens to be an exceptionally good researcher, writer and thinker. These conversations serendipitously dealt with a subject my friend had herself been actively researching and writing about, and it was her early insights, comments and encouragement which helped shape what I had originally intended to be one or two short articles into the present book.

Nikki's research and perspective in many ways parallels and strengthens what is covered in this book, and that is why I chose, with her permission, to include it here. In this series of exchanges Nikki describes her new-found climate change angst with a style and wit all her own, and if what has already been covered in this book has not sufficiently convinced you of the need for reform of the monetary system, just read what she has to say about what is coming if the increasingly centralized money power is not seriously challenged, restrained and ultimately vanquished.

Insights and perspectives from Nikki Alexander:

I DO have another article swirling around in my head – I hope it stops swirling soon so I can grab onto the structure and make sense out of it. I *know* something that is difficult to prove, rather than *believe* something hard-wired into my emotions. That's the core of the problem with trying to reason with people who have absorbed this climate change story.

The Chris Hedges speech on the death of the liberal class was so satisfying for me, so validating. (I'm not crazy.) He said something important (quoting Edward Bernays) – that people act based on imagery and emotion, not on knowledge. All it takes is an image of stranded polar bears, accompanied by string instruments, to make our tears flow.

Think of all the imagery (with music) we have been bombarded with, the violent hurricanes, earthquakes and floods, the magnificent animals driven into exile. Of course everyone with any shred of compassion and love for the earth will be mobilized to help. And this is what we are up against. I LOVE the earth, the beautiful plants and animals, all the teeny weeny things and the great big things that fit together in perfect harmony.

I never questioned the official narrative at all. It was just an accident that I stumbled into the truth. So how can you reach people with facts and knowledge when they are controlled by their emotional beliefs? (My mind just went to Galileo) To date $79 billion has been spent on public indoctrination – color brochures, DVDs, TV ads and shows, media blitz – all presented in colorful imagery that manipulates the emotions. If you dare to question this, people assume you don't love the earth and they hate you for *that*. But we DO love the earth! That's how they captured and recruited us. They know we love the earth.

Knowing this, and not having $79 billion at your disposal, how can you present your findings in such a way that people who love the earth will begin to doubt the official story or at least the official solutions? What idea will break through the trance of belief? Evoke imagery and stir emotions? Or perhaps showing how these official solutions will harm the earth even more – appeal to their existing belief. The official presentation is so slick and so convincing sometimes I wonder if I am wrong.

When I looked at the scale of reporting GHG data, how meticulously organized it is, and the $ trillions changing hands, I couldn't help but think – wouldn't this be wonderful if it were all directed to repairing the environmental destruction that is REAL... like the Gulf of Mexico, toxic sludge in Appalachia, contamination of Bhopal India, trash in the oceans, mercury in the water, GMO mutation of natural seeds etc. . . Clearly, they have enough money. The Carbon Disclosure Project (1) says they "act on behalf of 534 institutional investors, holding $64 trillion in assets under management."

This is just like the monetary system. No wonder we are focused on both of these issues. They are the same thing. It seems that no matter which door you open you find yourself in the same room…

. . . All of my friends are liberal progressive nature lovers who believe in climate change like a fundamentalist religion. They refuse to listen to ANYTHING that refutes the official narrative. As a result I have been ostracized as a 'climate denier' which makes me censor myself in their presence. I never questioned climate change at all and bought Al Gore's movie to become a better student. Something happened in [my community] that made me look at what is behind this and that led to more and more discoveries.

I could not initially understand why the "solutions" all involved massive transfers of wealth . . . and why "free" trade and Third World debt restructuring took up all the space in [official] documents. I kept waiting to come to a page about melting glaciers and stranded polar bears but not one word about nature appears in any of their documents. That's because this has nothing to do with nature. duh.

I realized that gradually, after reading the WTO, IMF, World Bank papers and tracing the 1992 Earth Summit backwards to Stockholm in 1972 where the real motive was stated in plain English – to replenish the dwindling supply of raw materials global bankers and traders depend on for their wealth and control.

Not only have they depleted their plantations in the Third World, the slaves themselves cannot possibly repay the multi-billion dollar debt forced on them by the IMF, Paris Club and transnational banks. Someone must pay for rehabilitating the plantation and someone must supply these parasites with a new debt-income stream. Maurice Strong opened the Rio Summit naming their target – "the affluent middle class in the industrialized countries."

Measuring and monetizing "carbon footprints" supplies the parasite casino with a global windfall profits scheme, while simultaneously consolidating external control over every country's energy, industry, transportation, agriculture, land use, housing, food and water – in perpetuity. This is the meaning of "sustainable." When they talk about sustainable development they mean sustainable revenue.

[According to information provided by The Carbon Disclosure Project] everything under the sun is being measured and monetized – from hospitals and schools to refrigerators, toothpaste and telephone calls – and entered into a global database to provide a matrix for calculating GLOBAL carbon fines. The website states that CDP is "the only global climate change reporting system" and "acts on behalf of 534 institutional investors, holding $64 trillion in assets under management and some 60 purchasing organizations." I was surprised by how transparent this is – who all their partners are and their board of trustees.

When you look at the SCALE of this operation it's mind boggling. There are trillions of dollars and thousands of organizations already engaged in this – right down to local municipalities that are signed up without their citizens' awareness. So the economy is not dead. This isn't a jobless recovery after all. Business is booming – if you're in the right business. . .

Global trade in carbon emissions is already [as of 2010] in the hundreds of billions. Point Carbon estimates they could grow to more than $1 trillion by 2020. Investors also trade derivatives contracts on the future price of carbon on exchanges.

EPA director, Lisa Jackson, was New Jersey's representative on the Regional Greenhouse Gas Initiative (RGGI), a cap-and-trade program for 10 Middle Atlantic and northeastern states.

The Chicago Climate Exchange was founded in 2001 with $1.1 million in seed money from the Joyce Foundation while Obama was on the Joyce Board. Richard L. Sandor, CCX chairman and chief executive, who is credited with inventing interest-rate futures, has attracted about 400 entities, including Monsanto, IBM, Intel, Bank of America, Citi and the Rockefeller Brothers Fund. Stuart Eizenstat, a CCX director, was Bill Clinton's lead climate treaty negotiator.

London-based Climate Exchange plc (CLE) owns the Chicago Climate Exchange (CCX), the Chicago Climate Futures Exchange (CCFX) and the European Climate Exchange (ECX) – Goldman Sachs owns 19% of Climate Exchange Plc (CLE.UK)

Climate Exchange (CLE) was acquired by Atlanta-based IntercontinentalExchange (ICE) in April 2010, giving them a virtual monopoly over Europe's estimated €100 billion carbon market and a significant lead in the US market.

ICE was established in 2000. It specializes in credit default swaps, derivatives, energy and commodity futures and carbon emissions. Its founding shareholders represent the world's largest energy companies and global financial institutions.

Its (for-profit) clearing house for credit default swaps, ICE US Trust LLC, was created in 2009 as a member of the Federal Reserve System with an exemption from the SEC and US Treasury. ICE US Trust and ICE Clear Europe have cleared $12 trillion in credit default swaps since their inception in 2009. ICE also operates ICE Futures Europe and ICE Futures US "providing integrated access to derivatives markets." Fred Hatfield, an ICE board director, is a member of Obama's Economic Policy Advisory Committee. Another board director, Sir Callum McCarthy, is a former Barclays Bank senior official and current Non-Executive Director of Her Majesty's Treasury. A third board director, Sir Robert Reid, is Chairman of ICE Futures Europe, ICE Clear Europe and a Deputy Governor of the Bank of Scotland.

ICE has quite an impressive list of acquisitions, including the International Petroleum Exchange (now ICE Futures Europe), the New York Board of Trade (now ICE Futures US), ChemConnect (natural gas), Winnipeg Commodity Exchange (now ICE Futures Canada), Chatham Energy, YellowJacket Software, Creditex (credit derivatives), TCC, a Commodity Futures Trading Commission, and has an exclusive licensing agreement with the Russell Investment Group. ICE is also partnered with TSX Group's Natural Gas Exchange.

I don't see any way to counteract this level of organized power – except to persuade people that trading carbon dioxide emissions will have no effect on the climate.

REFERENCES CITED

Chapter 1

1. http://www.nature.com/nature/journal/v403/n6767/full/403298a0.html *Halocarbons produced by natural oxidation processes during degradation of organic matter* by F. Keppler, R. Eiden, V. Niedan, J. Pracht & H. F. Schöler. Nature 403, 298-301 (January 20, 2000)
2. http://www.scienceagogo.com/news/19990904032112data_trunc_sys.shtml *The Role of Water Vapor in Climate Change* by Kate Melville. ScienceAGoGo, October 4, 1999. Accessed April 2013.
3. http://www.nasa.gov/topics/earth/features/vapor_warming.html *Water Vapor Confirmed As a Major Player in Climate Change,* by Kathryn Hansen. NASA's Goddard Space Flight Center. November 18, 2008. Accessed April 2013.
4. http://www.ncdc.noaa.gov/oa/climate/gases.html *Greenhouse Gasses: Frequently asked questions.* National Oceanic and Atmospheric Administration, Climate Data Center. Accessed March 2013.
5. http://www.enviroliteracy.org/article.php/428.html *Greenhouse Gases.* The Environmental Literacy Council. Accessed September 2011.
6. http://www.ehow.com/facts_6177559_nitrogen-gas-vs_-carbon-dioxide.html *Nitrogen Gas vs. Carbon Dioxide* by Eric Tilden, eHow Contributing Writer. Accessed September 2011.
7. http://www.guardian.co.uk/commentisfree/2008/aug/13/carbonemissions.climatechange *The Oxygen Crisis* by Peter Tatchell. Guardian.co.uk, August 13, 2008
8. http://nitrificationnetwork.org/Introduction.php Nitrification Primer, Oregon State University, c2004.

9. http://carbon-based-ghg.blogspot.com/2009/10/scientists-integrate-nitrogen-cycle.html *Scientists Integrate Nitrogen Cycle Into Climate Model,* posted by Brian Thomas. Carbon Based Climate Change Blogspot. October 10, 2009.
10. http://www.probeinternational.org/UPennCross.pdf *Global Warming Advocacy Science: a Cross Examination*, Research Paper No. 10-08. Jason Scott Johnston, University of Pennsylvania Institute for Law and Economics: A Joint Research Center of the Law School, the Wharton School, and the Department of Economics in the School of Arts and Sciences at the University of Pennsylvania. May 2010.
11. http://www.oism.org/pproject/s33p36.htm *Environmental Effects of Increased Atmospheric Carbon Dioxide.* Scholarly review by Arthur B. Robinson, Noah E. Robinson, and Willie Soon at the Oregon Institute of Science and Medicine. Accessed April 2013.
12. http://whatcom.wsu.edu/ag/compost/fundamentals/needs_carbon_nitrogen.htm *Compost Fundamentals*. Washington State University, Whatcom County Extension. Accessed September 2011.
13. http://www.sare.org/publications/bsbc/bsbc.pdf *Building Soils For Better Crops* by Fred Magdoff and Harold van Es. 3rd edition. Sustainable Agriculture Research and Education Program (SARE), 2009.
14. http://unitedstatesag.org/nitrogen.html *Why Do We Buy So Much Nitrogen?* U.S. Ag, LLC: Nature Friendly Agriculture. Accessed October 2011.
15. http://www.oasishd.ca/index.php?option=com_content&view=article&id=268:humus&catid=11:cath&Itemid=3 *Humus – Forgotten Climate Aide.* Oasis HD: Love Nature. Accessed April 2013.
16. http://www.gcrio.org/CONSEQUENCES/vol4no1/carbcycle.html *The Carbon Cycle, Climate, And The Long-Term Effects Of Fossil Fuel Burning* by James F. Kasting. Consequences:

The Nature and Implications of Environmental Change, Vol. 4, No. 1, 1998

17. http://www.epa.gov/climatechange/ghgemissions/gases/co2.html *Overview of Greenhouse Gases: Carbon Dioxide Emissions.* Environmental Protection Agency. Accessed April 2013.
18. http://www.ldeo.columbia.edu/res/div/ocp/drought/dust_storms.shtml *Did Dust Storms Make the Dust Bowl Drought Worse* by Ben Cook etal. Lamont-Doherty Earth Laboratory: The Earth Institute at Columbia University. C2011.
19. http://www.odl.state.ok.us/usinfo/maps/dustbowl/ *Is the Dust Bowl Returning?* US Government Information: Oklahoma Department of Libraries. C2012.
20. https://www.soils.org/story/2011/nov/sun/from-dust-bowl-to-dust-bowl-75-years-of-soil-science *From Dust Bowl to Dust Bowl: 75 Years of Soil Science.* Soil Science Society of America. C2013.
21. Per explanation provided by Ken Fousek, agricultural researcher, writer and contributor to http://sawsomme.org/
22. http://www.wiley.com/college/boyer/0470003790/reviews/pH/ph_water.htm *Water, PH and Non-covalent Bonding.* Concept Reviews. John Wiley & sons Publishers, c2002.
23. http://www.ucsusa.org/global_warming/science_and_impacts/impacts/the-impacts-of-land-use-on.html *The Impacts of Land Use on Climate Change*, Union of Concerned Scientists, July 14, 2003.
24. http://www.soilcarbon.com.au/case_studies/index.html Soil Carbon Case Studies. Soilcarbon.com, accessed September 2011.
25. http://www.globalagriculture.org/report-topics/climate-and-energy.html Agriculture at a Crossroads: Climate and Energy. Globalagriculture.org, accessed May 2013.
26. http://tinyurl.com/3dt82ou *The Effectiveness, Efficiency and Equity of Market-based and Voluntary Measures to Mitigate Greenhouse Gas Emissions from the Agri-food Sector* by Alexander Kasterine and David Vanzetti. Published in the

Trade and Environment Review 2010, United Nations Conference on Trade and Development (UNCTAD), Geneva.

27. Ibid
28. http://www.econexus.info/publication/carbon-market-dream *The Carbon Market Dream: Millions of Offsets from Land Use "Sinks."* Briefing for Cancun Climate Negotiations. November 2010. Econexus and Biofuelwatch. Accessed June 2013.
29. http://www.carbontradewatch.org/issues/carbon-offsets2.html *Carbon Offsets,* Carbon Trade Watch, accessed June 2013.
30. http://www.spiegel.de/international/europe/european-co2-emissions-reduction-system-is-broken-a-892134.html *CO2 Emissions: Can Europe Save Its Cap and Trade System* by Nils Klawitter. Spiegel Online International, April 3, 2013.
31. *The Origins of the Federal Reserve System: Money, Class and Corporate Capitalism, 1890-1913* by James Livingston. Cornell University Press, New York. 1986 and *Defying Corporations, Defining Democracy: A Book of History & Strategy,* edited by Dean Ritz. The Apex Press, New York, 2001. Also explore information provided by Lori Wallach, director of Public Citizen's Global Trade Watch: http://www.citizen.org/trade/article_redirect.cfm?ID=17010
32. http://www.oism.org/pproject/s33p36.htm *Environmental Effects of Increased Atmospheric Carbon Dioxide.* Oregon Institute of Science and Medicine.
33. http://www.sbs.com.au/news/article/1492651/Factbox-Carbon-taxes-around-the-world *Carbon Taxes Around the World.* Accessed April 2013.
34. http://www.guardian.co.uk/environment/graphic/2012/jul/17/geoengineering-world-map?intcmp=239 *Geoengineering Projects Around the World – Map.* By Finbarr Sheehy, Guardian Uk, July 17, 2012.

35. http://www.agriculturedefensecoalition.org/content/geoengineering-current-actions Agricultural Defense Coalition: Geoengineering Current Actions. Accessed August 2012.
36. http://viacampesina.org/en/index.php?option=com_content&view=article&id=287:the-myth-of-biofuels&catid=22:biodiversity-and-genetic-resources&Itemid=37 *The Myth of Biofuels* by Edivan Pinto etal. La Via Campesina, March 14, 2007.
37. http://www.wrm.org.uy/actors/CCC/sinks7.html *Climate Change Convention: Sinks that Stink.* World Rainforest Movement. Bulletins from 1998 through 2000.
38. http://www.wrm.org.uy/bulletin/155/carbon_trading.html *What Is Carbon Trading For?* By Chris Lang. World Rainforest Movement, Bulletin No. 155, June 2010.
39. http://www.emerging-markets.com/biodiesel/ *Biodiesel 2020: A Global Market survey,* 2nd edition. Emerging Markets Online: Global Energy and Biofuels Market Intelligence. Accessed September 2011.
40. http://www.ncdc.noaa.gov/oa/climate/gases.html *Greenhouse Gases: Frequently Asked Questions.* National Oceanic and Atmospheric Administration, National Climatic Data Center. Accessed September 2011.
41. http://www.i-sis.org.uk/Why_Glyphosate_Should_be_Banned.php *Why Glyphosate Should Be Banned.* Institute of Science in Society Special Report 10/10/2012. Accessed April 2013.
42. http://www.wrm.org.uy/bulletin/39/research4.html *Forests better than plantations, even as carbon sinks.* World Rainforest Movement, Bulletin No. 39, October 2000
43. http://www.non-gmoreport.com/articles/may10/consequenceso_widespread_glyphosate_use.php *Scientist warns of dire consequences with widespread use of glyphosate.* Interview appearing in the Organic NonGMO Report, December/January 2010.

44. http://pubs.acs.org/doi/abs/10.1021/tx1001749 *Glyphosate-Based Herbicides Produce Teratogenic Effects on Vertebrates by Impairing Retinoic Acid Signaling* by Alejandra Paganelli, etal. Chem. Res. Toxicol., 2010, 23(10), pp 1586–1595. Publication Date (Web): August 9, 2010.
45. http://www.wrm.org.uy/actors/CCC/sinks7.html *Climate Change Convention: Sinks that Stink.* World Rainforest Movement. Bulletin, November, 1999.
46. http://www.fao.org/docrep/w5183e/w5183e06.htm *The CO2 fertilization effect: higher carbohydrate production and retention as biomass and seed yield.* L. Hartwell Allen, Jr., Jeff T. Baker, and Ken J. Boote. FAO Corporate Document Repository.
47. http://www.gcrio.org/CONSEQUENCES/vol4no1/carbcycle.html The C*arbon Cycle, Climate, And The Long-Term Effects Of Fossil Fuel Burning* by James f. Kasting. Consequences: The Nature and Implications of Environmental Change, Vol. 4, No. 1, 1998
48. http://www.co2science.org/education/truthalerts/v13/cowpea.php Seeing Is *Believing*, video clip. Accessed September 2011.
49. http://www.omafra.gov.on.ca/english/crops/facts/00-077.htm *Carbon Dioxide in Greenhouses,* Factsheet. Ontario Ministry of Agriculture, Food and Rural Affairs. Dec. 2002
50. http://carboncyclesandsinks.org/2010/06/agricultural-intensification-cuts-greennhouse-emissions/ *Agricultural Intensification "Cuts Greenhouse Emissions"* by Richard Douthwaite, June 29, 2010. Carbon Cycles and Sinks Network: Developing Policy for land-based carbon sequestration in Ireland.
51. http://www.nutritionsecurity.org/PDF/NSI_White%20Paper_Web.pdf *Human Health, the Nutritional Quality of Harvested Food and Sustainable Farming Systems* by John B. Marler and Jeanne R. Wallin. Nutrition Security Institute, 2006.
52. http://www.mineralresourcesint.co.uk/pdf/mineral_deplet.pdf *A study on the mineral depletion of the foods available to us as a nation over the period 1940 to 1991.* The data used as the basis

for this study was published in 5 Editions, initially under the auspices of the Medical Research Council and later the Ministry of Agriculture Fisheries and Foods and the Royal Society of Chemistry: Authors R.A. McCance and E.M. Widdowson. Copyright, David E. Thomas, January 2000.

53. http://www.soilandhealth.org/01aglibrary/010119lowdermilk.usda/cls.html *Conquest of the Land Through Seven Thousand Years* by W. C. Lowdermilk, USDA, Soil Conservation Service, February 1948.
54. http://www.radford.edu/~wkovarik/envhist/1ancient.html *Environmental History Timelines: Ancient Civilizations.* Radford University. Accessed September 2011.
55. http://www.environmentalgraffiti.com/offbeat-news/environmentalism-in-3000-bc/912 *Environmentalism in 3000B.C.* By Chris Ingham Brooke. Environmental Graffitti On Line. Accessed September 2011.
56. http://rewild.info/anthropik/2007/05/answering-the-archdruid/index.html *Archdruid Watch: Glimpsing the Deindustrial Age* by Jason Godesky. May 30, 2007.
57. http://news.bbc.co.uk/2/hi/science/nature/722907.stm *Pollution Damages Intelligence* by David Kirby. BBC News, April 22, 2000.
58. http://pielkeclimatesci.wordpress.com/2010/02/26/our-paper-impacts-of-land-use-land-cover-change-on-climate-and-future-research-priorities-by-mahmood-et-al-2010-as-appeared/ *Impacts Of Land Use Land Cover Change On Climate And Future Research Priorities.* By Mahmood Et Al. First published in 2009.
59. http://www.newsdesk.umd.edu/scitech/release.cfm?ArticleID=761 News Release: Land Use May Be Causing Half of Earth's Surface Warming, study published in the Journal Nature by Eugenia Kalnay and Ming Cai, scientists in the University of Maryland's Department of Meteorology. May 28, 2003.

60. http://www.ucsusa.org/global_warming/science_and_impacts/impacts/the-impacts-of-land-use-on.html The Impacts of Land Use on Climate Change, Union of Concerned Scientists, July 14, 2003.
61. http://e360.yale.edu/content/feature.msp?id=2196 *The Other Inconvenient Truth: The Crisis in Global Land Use* by Jonathan Foley.

Chapter 2

1. http://www.pollutionissues.com/Re-Sy/Soil-Pollution.html#ixzz0w85zMIhE *Soil Pollution.* Pollution Issues, accessed September 2011.
2. http://www.epa.gov/airtrends/ *Air Trends*, U.S. Environmental Protection Agency. Accessed September 2011.
3. http://www.policyalmanac.org/environment/archive/acid_rain.shtml *What Is Acid Rain and What Causes It?* Adapted from "Acid Rain", Environmental Protection Agency. August 6th, 2002. Almanac of Policy Issues, accessed September 2011.
4. http://www.usgs.gov/newsroom/article.asp?ID=3078 *Acid Rain Study Show Substantial Decreases But More Progress Is Needed.* U.S. Geological Survey, released January 19, 2012.
5. http://www.scorecard.org/env-releases/def/ej_burdens.html *Pollution Locator: Environmental Burdens Scorecard.* The Pollution Information Site. Accessed September 2011.
6. http://www.epa.gov/climatechange/endangerment.html Endangerment and Cause or Contribute Findings for Greenhouse Gases under Section 202(a) of the Clean Air Act. Signed by EPA Administrator on December 7, 2009. Accessed September 2011.
7. http://www.epa.gov/air/ozonepollution/ *Ground Level Ozone*, U.S. Environmental Protection Agency. Accessed September 2011.

8. http://www.atmosphere.mpg.de/enid/Information 2/Ozone smog 5hw.html *Tropospheric Ozone*, Quantify. Accessed September 2011.
9. http://www.nasa.gov/topics/earth/features/nox lightning.html *NASA Researchers Explore Lightning's NOx-ious Impact on Pollution*, Climate. NASA, 10-22-09.
10. http://www.plantstress.com/Articles/Oxidative%20Stress.htm *Oxidative Stress*, by Bryan D. McKersie. Dept of Crop Science, University of Guelph, December, 1996.
11. http://www.accessdata.fda.gov/scripts/cdrh/cfdocs/cfcfr/CFRSearch.cfm?fr=801.415 Code of Regulations, 21CFR801.415. U.S. Food and Drug Administration. Accessed September 2011.
12. http://www.newworldencyclopedia.org/entry/Ozone *Ozone*. New World Encyclopedia. Accessed September 2011.
13. http://www.ozonetherapy.co.uk/articles/ed mccabe ozone history and references.htm *Scientific And Medical References Proving Ozone's Validity As A Medical Treatment* by Ed McCabe. Accessed September 2011.
14. http://www.meridianinstitute.com/eaem/thermo/emtherm1.htm *The Thermo-Ozone Generator: Its Discovery, Philosophy of Action, Principle of Cure, with Directions for the Cure of Disease, A New discovery in Medicine* by S. R. Beckwith, M.D. 1897. Meridian Institute. Accessed September 2011.
15. http://www2.canada.com/components/print.aspx?id=5a989e08-ff58-4b6f-96d5-f1dfd3fbd909 *Let's Clear the Air on Ozone Dangers*, by Joel Schwarz. Canada.Com, July 5, 2007.
16. http://www.johnlocke.org/acrobat/spotlights/02 ozone & asthma.pdf *Check The Facts Next Time: Ozone-Asthma Link Has No Foundation* by Dr. Roy Cordato. Spotlight No. 226, October 16, 2002.
17. http://wayback.archive.org/web/*/http://www.atsdr.cdc.gov/document/ASTHMA BRONX FINAL REPORT.pdf *A Study of Ambient Air Contaminants and Asthma in New York City. Part A and B, Final Report*, July 2006. Prepared by the

New York State Department of Health for The U.S. Department of Health and Human Services Agency for Toxic Substances and Disease Registry. Web Archived. Accessed September 2011.

18. http://www.webmd.com/asthma/guide/asthma-triggers *Asthma Causes and Triggers,* WebMD.com. Accessed September 2011.
19. http://wiki.answers.com/Q/What_is_the_major_chemical_found_in_smog *What Is the Major Chemical Found in Smog?* Answers.com, Accessed September 2011.
20. http://www.epa.gov/cleanenergy/energy-and-you/glossary.html *Clean Energy: Glossary.* U.S. Environmental Protection Agency. Accessed September 2011.
21. http://web.archive.org/web/20081225215756/http://www.environment-agency.gov.uk/cy/busnes/pynciau/llygredd/39229.aspx? *Sulphur Dioxide.* Environment Agency, Wales. Web archived. Accessed September 2011.
22. http://toxtown.nlm.nih.gov/text_version/chemicals.php?id=29 *Sulphur Dioxide.* Tox Town, National Library of Medicine. Accessed September 2011.
23. http://www.epa.gov/climatechange/glossary.html *Glossary of Climate Change Terms,* U.S. Environmental Protection Agency. Accessed September 2011.
24. http://www.greenfacts.org/en/ozone-o3/index.htm *Scientific Facts on Air Pollution.* GreenFacts Digest. Accessed September 2011.
25. http://en.wikipedia.org/wiki/Nitrogen_dioxide *Nitrogen Dioxide,* Wikipedia: The Free Encyclopedia. Accessed September 2011.
26. http://www.epa.gov/climatechange/glossary.html *Glossary of Climate Change Terms,* U.S. Environmental Protection Agency. Accessed September 2011.
27. http://www.nature.com/ngeo/journal/v2/n11/full/ngeo678.html *Atmospheric science:Nitrous oxide delays ozone recovery* by

Martyn Chipperfield. Nature Geoscience: 2, 742 – 743 (2009)

28. http://www.nofa.org/tnf/nitrogen.php *Ecologically Sound Nitrogen Management: One Researcher's Educated Guess* by Mark Schonbeck. NOFA on line. Accessed 2011.
29. http://home.cc.umanitoba.ca/~vsmil/pdf_pubs/Nitrogen%20and%20Food%20Production.pdf *Nitrogen and Food Production: Proteins for Human Diets* by Vaclav Smil. Ambio Vol. 31 No. 2, March 2002
30. http://www.guardian.co.uk/commentisfree/2008/aug/13/carbonemissions.climatechange *The Oxygen Crisis* by Peter Tatchell. Guardian.co.uk, August 13, 2008.
31. http://academic.research.microsoft.com/Paper/12561806.aspx *The Myth of Nitrogen Fertilization for Soil Carbon Sequestration*, S. A. Khan, R. L. Mulvaney, T. R. Ellsworth and C. W. Boast. Published in Journal of Environmental Quality 36:1821-1832 (2007)
32. http://www.grist.org/article/2010-02-23-new-research-synthetic-nitrogen-destroys-soil-carbon-undermines-/ *The Soil Is Bleeding* by Tom Philpott. GRIST, 23 FEB 2010
33. http://www.ars.usda.gov/research/publications/publications.htm?SEQ_NO_115=222066 *Soil Carbon Sequestration to Mitigate Climate Change and Advance Food Security productivity*, by Ronald Follett etal. Agricultural Research Service, USDA. Published in Soil Science 172 (12): 943-956, Dec. 28, 2007.
34. http://www.broowaha.com/articles/6118/all-of-our-aquifers-are-leaking *All of Our Aquifers Are Leaking* by Tom Lewis (editor), BrooWaha, Your Shared News and Views. February 01, 2010
35. http://www.wateryear2003.org/en/ev.php-URL_ID=1607&URL_DO=DO_TOPIC&URL_SECTION=201.html *Facts and Figures: The Different Water Users.* International Year of Freshwater 2003. Accessed September 2011.

36. http://findarticles.com/p/articles/mi_7476/is_201001/ai_n52374745/ *Impacts of Land Use/Land cover Change on Climate and Future Research Priorities*, by Rezau Mahmood etal. Bulletin of the American Meteorological Society, January 2010.
37. http://www.globalchange.umich.edu/globalchange2/current/lectures/freshwater_supply/freshwater.html *Human Appropriation of the World's Fresh Water Supply.* Global Change, January 4, 2006. Website material copyright University of Michigan. Accessed September 2011.
38. http://ga.water.usgs.gov/edu/wuir.html *Irrigation Water Use: Water Science for Schools.* U.S. Geological Survey, Accessed September 2011.
39. http://www.tufts.edu/programs/feast/animalwelfare.htm *Feast, Food and Animal Welfare.* Tufts University. Accessed September 2011.
40. http://www.time.com/time/magazine/pdf/100125_feed_vs_pastures.pdf *Feedlots vs Pastures.* Time.com. Accessed April 2013.
41. http://tinyurl.com/2bfyd75 *Abstract: Methane Emissions from Swine Houses in North Carolina* by R.R. Sharpe, L.A. Harper, J.D. Simmons. Chemosphere – Global Change Science. Volume 3, Issue 1, January 2001, Pages 1-6
42. http://findarticles.com/p/articles/mi_m0ASV/is_1_24/ai_54466913/ *CAFOs Feed a Growing Problem* by Kevin Stubbs and Karen Cathey, Endangered Species Bulletin, January 1999.
43. http://www.sciencedaily.com/releases/2008/04/080421143836.htm *Fertilizer Run-Off From Agricultural Activities Blamed For Gulf Dead Zone In Gulf Of Mexico.* Science Daily, April 24, 2008.
44. http://www.biofuelwatch.org.uk/ BiofuelWatch, 2006.
45. http://www.news-medical.net/news/20090623/102-billion-starving-people-worldwide-UN-says.aspx *1.02 Billion Starving Worldwide, U.N. Says.* The Medical News, June 23, 2009.

46. http://www.fao.org/english/newsroom/news/2002/9703-en.html *What the New Figures on Hunger Mean.* Food and Agriculture Organization of the United Nations. FAO, 2002.

Chapter 3

1. http://www.soilcarbon.com.au/case studies/index.html *Soil Carbon Case Studies*, Soil Carbon (Australia) Pty Ltd, 2007. Accessed September 2011.
2. http://www.newsdesk.umd.edu/scitech/release.cfm?ArticleID=761 News Release: Land Use May Be Causing Half of Earth's Surface Warming, study published in the Journal Nature by Eugenia Kalnay and Ming Cai, scientists in the University of Maryland's Department of Meteorology. May 28, 2003.
3. http://web.archive.org/web/20081202075528/http://www.familyfarmdefenders.org/pmwiki.php/Agrofuels/TheThreatOfAgrofuels *The Threat of Agrofuels* by John Peck. Family Farm Defenders, May 25, 2008. Web Archived.
4. http://www.news-medical.net/news/20090623/102-billion-starving-people-worldwide-UN-says.aspx *1.02 Billion Starving Worldwide, U.N. Says.* The Medical News, June 23, 2009.
5. http://www.fao.org/english/newsroom/news/2002/9703-en.html *What the New Figures on Hunger Mean.* Food and Agriculture Organization of the United Nations. FAO, 2002.
6. http://www.ewg.org/agmag/2010/05/farm-income-data-debunks-subsidy-myths/ *Farm Income Data Debunks Subsidy Myths,* by David DeGennaro, May 12, 2010.
7. http://www.downsizinggovernment.org/agriculture/subsidies *Agricultural Subsidies* by Chris Edwards. Downsizing the Federal Government: Cato Institute, June 2009.
8. http://www.countercurrents.org/glo-shiva050404.htm *The Suicide Economy Of Corporate Globalisation* by Vandana Shiva, 05 April, 2004, Znet. Accessed September 2011.

9. http://www.foodfirst.org/node/239 T*he Myth – Scarcity: The Reality – There IS Enough Food* by Melissa Moore. Food First: Institute for Food and Development Policy, February 8, 2005. Accessed September 2011.
10. http://www.harpers.org/archive/2004/02/0079915 *The Oil We Eat: Following the Food Chain Back to Iraq* by Richard Manning, Harper's Magazine Archives, February 2004. Accessed September 2011.
11. http://en.wikipedia.org/wiki/Wetland *Wetland*, from WIKIPEDIA, The Free Encyclopedia, accessed September 2011.
12. http://tinyurl.com/2adde4e Lecture by Dr. Tom Sever, sole archaeologist for NASA's Marshall Space Flight Center. accessed September 2011.
13. http://www.forestinfo.org/sites/default/files/pdfReports/11-3sfii.pdf *Sustainable Forestry II*, Econ-Link, Linking Social, Economic and Ecological Issues. Volume 11, Number 3. Last accessed July 2013.
14. http://www.wetlands.org/Whatwedo/PeatlandsandCO2emissions/Aboutpeatlands/tabid/1362/Default.aspx *About Peatlands.* Wetland International, Wetlands for Water and for Life. Accessed September 2011.
15. http://www.savetherainforest.org/savetherainforest_006.htm *Causes of Rainforest Destruction*, Save The Rainforest, c2001-2005.
16. http://peat-portal.net/index.cfm?&menuid=123&parentid=113 *Assessment on Peatlands, Biodiversity and Climate change: Executive Summary*. Peat-Portal. Retrieved August 2, 2010.
17. ibid
18. http://rainforests.mongabay.com/20indonesia.htm "Tropical Rainforests: Indonesia." by Butler, Rhett A. Mongabay: A Place Out of Time: Tropical Rainforests and the Perils They Face. February 6 2006.

19. http://www.rain-tree.com/facts.htm *Rainforest Facts, the Disappearing Rainforests.* Copyright 1997 to present, Raintree Nutrition Inc. Last updated 3-20-2010.
20. http://www.savetherainforest.org/savetherainforest_006.htm *Causes of Rainforest Destruction.* Save The Rainforest, c2001-2005.
21. http://web.archive.org/web/20100828195734/http://www.truth-out.org/how-climate-change-south-america-affects-entire-planet-an-interview-with-nikolas-kozloff61179 *How Climate Change in South America Affects the Entire Planet: An Interview With Nikolas Kozloff,* by Joshua Frank. T r u t h o u t | Interview, Wednesday 14 July 2010.
22. http://www.wrm.org.uy/actors/CCC/sinks7.html *Climate Change Convention: Sinks that Stink.* World Rainforest Movement. Bulletin 2000.
23. http://www.harpers.org/archive/2004/02/0079915 *The Oil We Eat: Following the Food Chain Back to Iraq* by Richard Manning, Harper's Magazine: Archives. February 2004.
24. http://www.grain.org/seedling/?id=682 *Watershed Cattle* by John Wilson, GRAIN, April 2010.
25. http://ag.arizona.edu/OALS/ALN/aln49/olsson.html *The Potential Benefits of Carbon Sinks in Dryland Agricultural Soils* by Lennart Olsson, Andrew Warren, and Jonas Ardö. Arid Lands Newsletter, No. 49, May/June 2001.
26. http://www.wrm.org.uy/bulletin/126/Zimbabwe.html *Zimbabwe: Tree plantations to accrue greenhouse gases emissions credits* . World Rainforest Movement, Bulletin No. 126, January 2008.
27. http://wrm.org.uy/bulletin/158/Africa.html *The Plunder of Africa Continues.* World Rainforest Movement, 2010 update. Accessed June 2013.
28. http://grazingmanagement.blogspot.com/ *Grazing Management for Healthy Soils* by Dr. Christine Jones. Posted on the Grazing Management Blogspot by Michael Kiley, March 24, 2006.

29. http://www.eatwild.com/environment.html Grazed pasture is the best land use for storing carbon, Report posted at bottom of Environment page of the Eat Wild website. Accessed October 2012.
30. http://www.greenmoneyjournal.com/article.mpl?newsletterid=41&articleid=549 *Moving Our World Towards Sustainability* by Allan Savory and Christopher Peck. Green Money Journal. Vol. 18, Issue 4, Number 76, Summer 2010.
31. http://www.ewg.org/agmag/2011/06/city-slickers-continue-to-rake-in-farm-payments/ *City Slickers Continue To Rake In Farm Payments.* Posted by Sara Sciammacco in Subsidies on June 23, 2011. Environmental Working Group: Agriculture.
32. http://www.farmersweekly.co.za/index.php?p[IGcms_nodes][IGcms_nodesUID]=f184f29916e50de7f6ab5978f7392278 *Going "Beyond Organic" Makes Small Farms Profitable* by Robyn Joubert. Farmers Weekly. January 6, 2009.
33. http://www.guardian.co.uk/lifeandstyle/2010/jan/31/food-industry-environment *Interview: Joel Salatin* by Gaby Wood. The Observer, January 31, 2010.
34. http://www.americanchronicle.com/articles/view/57222 *America's Silent Killing Fields* by Barbara Peterson. The American Chronicle, April 16, 2008.
35. http://www.safefoodinc.org/index.php?option=com_content&view=article&id=3&Itemid=11 *About Food, Inc: Myths and Facts.* SafeFoodInc.com, copyright, 2009.
36. http://www.farmsanctuary.org/issues/factoryfarming/poultry/ *Factory Poultry Production.* Farm Sanctuary Inc. Accessed September 2011.
37. http://www.navdanya.org/climate-change/agriculture Climate Change & Agriculture, (c) 2009 Navdanya Trust. Accessed October 2012.t
38. http://www.perennialsolutions.org/carbon-sequestering-agriculture-global-warming-solution-piece-remove-co2-from-atmosphere-organic-garden.html Carbon Sequestering

Agriculture. By Eric Toensmeier. Accessed June 2013.co2-from-atmosphere-organic-garden.html

39. http://www.worldbank.org/html/cgiar/press/news0102.htm *New Study Reveal That Environmental Damage Threatens Future World Food Production: News Release.* Michael Rubinstein, Adlai J. Amor, Shirley Geer. February 24, 2001. Accessed September 2011.

Chapter 4

1. http://www.guardian.co.uk/business/2008/nov/07/bp-renewable-energy-oil-wind *Blow to Brown as BP scraps British renewables plan to focus on US*, by Terry Macalister. The Guardian, Thursday 6 November 2008
2. http://english.aljazeera.net/indepth/opinion/2011/06/20116711756667987.html *The great land grab: India's war on farmers*, by Vandana Shiva. Aljazeera, June 7, 2011.
3. http://www.guardian.co.uk/world/2011/jun/08/us-universities-africa-land-grab *US universities in Africa 'land grab'* by John Vidal and Claire Provost. The Guardian, June 8, 2011.
4. http://www.biofuelwatch.org.uk/ Biofuel Watch, home page. Accessed September 2011.
5. http://mountainjustice.org/facts/steps.php *What is Mountaintop Removal Mining.* Mountain Justice Website, accessed September 2011.
6. http://www.insidevandy.com/drupal/node/10620 *Some Problems with Solar* by Thomas Shattuck. Inside Vandy, September 13, 2009.
7. http://spectrum.ieee.org/green-tech/geothermal-and-tidal/earthquakes-hinder-green-energy-plans *Earthquakes Hinder Green Energy Plans* by Peter Fairley. IEEE Spectrum: Inside Technology. April 2011.

8. http://www.internationalrivers.org/node/3845 *Earthquakes Triggered by Dams.* International Rivers, accessed September 14, 2011.
9. http://blogs.smithsonianmag.com/science/2009/02/a-link-between-dams-and-earthquakes/ *A Link Between Earthquakes and Dams?* Posted by Sarah Zielinski. Smithsonian: Surprising Science. February 3, 2009.
10. http://spectrum.ieee.org/energy/environment/the-water-cost-of-carbon-capture *The Water Cost of Carbon Capture* by Samuel K. Moore, IEEE Spectrum: Inside Technology, June 2010.
11. http://www.globalchange.umich.edu/globalchange2/current/lectures/freshwater_supply/freshwater.html *The Water Resources of the Earth*, January 4, 2006. Global Change: University of Michigan, copyright 2000.
12. http://online.wsj.com/article/SB10001424052702303936704576398462932810874.html *The Facts about Fracking*, Wall Street Journal, June 25, 2011.
13. http://www.livablefutureblog.com/2011/10/fossil-energy-alternates *Fossil Energy Alternates: Shale Gas, Oil Shale, and Tar Sands*, by Dennis Keeney, visiting scholar: Center for a Livable Future. October 21, 2011. Accessed April 2013.
14. http://www.sourcewatch.org/index.php?title=Natural_gas *Natural gas as an alternative to coal.* Source Watch, accessed September 2011.
15. http://www.hydraulicfracturing.com/Water-Usage/Pages/Information.aspx *Hydraulic Fracturing Facts: Water Usage*, Hydraulic Fracturing Website sponsored by the Chesapeake Energy Corporation, Accessed September 14, 2011.
16. http://www.scientificamerican.com/article.cfm?id=shale-gas-and-hydraulic-fracturing *What the Frack? Natural Gas from Subterranean Shale Promises U.S. Energy Independence – With Environmental Costs* by David Biello. Scientific American, March 30, 2010.

17. http://www.time.com/time/video/player/0,32068,876880045001_2062814,00.html *The Fuss Over Fracking: The Dilemma of a New Gas Boom.* Video. Time online. April, 2013.
18. http://www.water-contamination-from-shale.com/main/fracking-study-sees-public-health-risks/ *Fracking Study Sees Public Health Risks.* Water Contamination from Shale website, accessed October 2011.
19. http://www.foreffectivegov.org/node/10353 *EPA Finds Secret Fracking Chemicals In Drinking Water* by Brian Turnbaugh. Center For Effective Government, August 31, 2009.
20. http://www.foxnews.com/scitech/2011/03/01/fracking-earthquakes-arkansas-man-experts-warn/ *Earthquakes in Arkansas May Be Man-Made, Experts Warn,* by Alec Liu and Jeremy a. Kaplan. FoxNews.com, March 1, 2011.
21. http://www.thestar.com/business/article/1000638 – shale-gas-fracking-halted-in-uk-amid-earthquake-fears *Shale gas 'fracking' halted in UK amid earthquake fears,* by Jill Lawless. thestar.com, June 1, 2011.
22. http://www.cbc.ca/news/technology/story/2012/04/17/environment-fracking-earthquake-studies.html *Fracking Causes Earthquakes, studies confirm.* CBC News: technology and Science. April 17, 2012.
23. http://www.nei.org/Key-Issues/protectingtheenvironment Key Issues: Protecting the Environment. Nuclear Energy Institute Website, accessed July 2013.
24. http://www.fair.org/index.php?page=3258 *Money Is the Real Green Power: The hoax of eco-friendly nuclear energy* by Karl Grossman. Fairness and Accuracy in Reporting EXTRA! January/February 2008.
25. http://eandt.theiet.org/magazine/2011/04/nuclear-perception.cfm *Nuclear Power After Fukushima* by Mark Venables. Energy and Technology Magazine, April 15, 2011.
26. http://www.altenergy.org/renewables/solar.html Solar Energy: Alternative Energy, accessed October 2011.

27. http://www.insidevandy.com/drupal/node/10620 *Some Problems with Solar*, by Thomas Shattuck. Inside Vandy, September 13, 2009.
28. http://www.savewesternny.org/news/The_hidden_cost_of_wind_turbines.html *The Hidden Cost of Wind Turbines* by Ed Douglass. Save Western New York, July 8, 2006.
29. http://www.iberica2000.org/es/Articulo.asp?Id=1457 *Windfarms Negative – Contamination of Aquifers and Ground Water*. Iberica2000, accessed October 2011.
30. http://www.aweo.org/problemwithwind.html *A Problem with Wind Power*, by Eric Rosenbloom. AWEO, Accessed October 2011.
31. http://www.plosone.org/article/info:doi/10.1371/journal.pone.0006802 *Energy Sprawl or Energy Efficiency: Climate Policy Impacts on Natural Habitat for the United States of America* by Robert I McDonald etal. PLoS ONE 4(8): e6802. doi:10.1371/journal.pone.0006802, c2009
32. http://www.americanhunter.org/ArticlePage.aspx?id=2843&cid=56 *Of Windmills and Wildlife*, By Frank Miniter, Executive Field Editor. American Hunter, July 14, 2010.
33. http://alternativeenergy.procon.org/view.answers.php?questionID=001244 *Can Alternative Energy Effectively Replace Fossil Fuel?* Alternative Energy ProCon.org, Accessed October 2011.
34. http://spectrum.ieee.org/energy/fossil-fuels/joules-btus-quadslets-call-the-whole-thing-off *Joules, BTUs, Quads – Let's Call the Whole Thing Off* by Harry Goldstein, William Sweet. IEEE Spectrum: Inside Technology. January 2007.
35. http://www.radford.edu/~wkovarik/oil/3unconventional.html *The Oil Reserve Fallacy: Proven Reserves Are Not a Measure of Future Supply*, by Dr. Bill Kovaric, c2003. Posted at Radford University, accessed October 2011.
36. http://blogs.the-american-interest.com/wrm/2012/07/15/energy-revolution-2-a-post-post-american-post/ *Energy Revolution 2: a Post Post-American*

Post by Walter Russell Mead, July 15, 2012.The American Interest online, accessed November 2012.

37. http://www.nytimes.com/2008/02/08/science/earth/08wbiofuels.html?pagewanted=1 *Biofuels Deemed a Greenhouse Threat* by Elisabeth Rosenthal. The New York Times, February 8, 2008.
38. http://www.timberwatch.org.za/index.php?id=40 *Pertinent Issues*: Timberwatch coalition. Accessed September 2011.
39. http://www.timberwatch.org.za/index.php?id=41 *Agrofuels Page*: Timberwatch Coalition. Accessed September 2011.
40. http://www.energyfuturecoalition.org/biofuels/index.htm *The Biofuels Facts Page:* Energy Future Coalition (A United Nations Foundation). Accessed September 2011.
41. http://www.foodfirst.org/en/node/2426 *Agrofuels in the Americas,* by Richard Jonasse. Food First: Institute for Food and Development Policy, April 27, 2009.
42. http://e360.yale.edu/feature/the_case_against_biofuels_probing_ethanols_hidden_costs/2251/ *The Case Against Biofuels: Probing Ethanol's Hidden Costs*, by C. Ford Runge. Yale Environment 360, March 11, 2010.
43. http://ipsnews.net/news.asp?idnews=39740 *Biofuels – Great Green Hope or Swindle* by Stephen Leahy. Inter Press Service, October 20, 2007.
44. http://web.archive.org/web/20081202075528/http://www.familyfarmdefenders.org/pmwiki.php/Agrofuels/TheThreatOfAgrofuels *The Threat of Agrofuels by*: John E. Peck Executive Director, Family Farm Defenders. Archived. Accessed October 2011.
45. http://web.archive.org/web/20081228211258/http://www.familyfarmdefenders.org/pmwiki.php/S/AgrofuelMoratoriumCall *Agrofuel Moratorium Call,* Family Farm Defenders. Archived. Accessed October 2011.
46. http://www.globalsecurity.org/military/library/report/call/call_10-10-ch01.htm *Agribusiness Development Teams* (ADT) in

Afghanistan Handbook, November 2009. Global Security website, accessed October 2011..

47. http://findarticles.com/p/articles/mi_7476/is_201001/ai_n52374745/ *Impacts Of Land Use Land Cover Change On Climate And Future Research Priorities.* By Mahmood Et Al. 2009
48. http://www.newsweek.com/2009/09/18/big-oil-goes-green-for-real.html *Big Oil Goes Green For Real*, by Rana Foroohar. Newsweek, September 18, 2009.
49. http://www.innovativeindustry.net/biofuel-investments-of-big-oil-companies-part1 *Biofuel Investments of Big Oil Companies Part 1* by admin. Innovative Industry, January 4, 2011.
50. http://www.innovativeindustry.net/biofuel-investments-of-major-oil-companies *Biofuel Investments of Major Oil Companies Part 2* by admin. Innovative Industry, January 11, 2011.
51. http://www.grist.org/article/first-came-superweeds-and-now-come-the-superbugs *First Came Superweeds, Now Comes Superbugs* by Tom Laskawy. Grist, March 24, 2010.
52. http://responsibletechnology.org/docs/140.pdf Brochure from Responsible Technology's Tipping Point Network. Accessed July 2013.
53. http://www.grain.org/article/entries/598-corporate-power-agrofuels-and-the-expansion-of-agribusiness *Corporate Power: Agrofuels and the Expansion of Agribusiness.* GRAIN, July 24, 2007.
54. http://www.mdpi.com/1996-1073/1/1/35/ *Corn and Cellulosic Ethanol Cause Major Problems* by David Pimentel and Marcia Pimentel. Energies 2008, 1(1), 35-37; doi:10.3390/en1010035
55. http://www.scientificamerican.com/article.cfm?id=algae-biofuel-growth-environmental-impact *Is Algae Worse Than Corn for Biofuels?* By Katie Howell. Scientific American, January 22, 2010.

56. http://www.stockhouse.com/columnists/2008/june/11/algae-biofuel-mania *Algae Biofuel Mania* by Dan Sweeney. Stockhouse, June 11, 2008.
57. http://www.lowtechmagazine.com/2008/04/algae-fuel-biof.html *Leave the Algae Alone*, Low Tech Magazine, April 4, 2008.

Chapter 5

1. http://en.wikipedia.org/wiki/National_oil_company *National Oil Company* (article), Wikipedia. Accessed September 14, 2011.
2. http://www.innovativeindustry.net/biofuel-investments-of-major-oil-companies *Biofuels Investments of Major Oil Companies, Part 2.* Innovative Industries, January 11, 2011.
3. http://www.bloomberg.com/apps/news?pid=newsarchive&sid=aXRBOxU5KT5M *Carbon Capitalists Warming to Climate Market Using Derivatives*, by Lisa Kassenaar. Bloomberg, December 4, 2009.
4. http://news.bbc.co.uk/2/hi/science/nature/8343489.stm *Harrabin's Notes: Carbon trading.* BBC News, November 5, 2009.
5. http://www.zerohedge.com/article/cap-and-trade-gigantic-scam *Cap and Trade: A Gigantic Scam*, posted on Zerohedge from Washington's Blog, May 22, 2010.
6. http://www.washingtonsblog.com/2009/12/woman-who-invented-credit-default-swaps.html *Woman Who Invented Credit Default Swaps is One of the Key Architects of Carbon Derivatives, Which Would Be at the Very CENTER of Cap and Trade.* Posted on Washington's Blog, December 8, 2009.
7. http://www.platts.com/RSSFeedDetailedNews/RSSFeed/ElectricPower/8429015 *Oversight sufficient for US carbon derivatives market: report*, by Geoffrey Craig. Posted on Washington Platts, January 19, 2011.

8. http://www.thetwofacesofmoney.com/index.php/Site/SlidePresentation *Money Owed and Owned*, Slide presentation by Geraldine Perry and Ken Fousek.
9. http://www.grain.org/article/entries/598-corporate-power-agrofuels-and-the-expansion-of-agribusiness *Corporate power: Agrofuels and the expansion of agribusiness.* Grain, July 24, 2007.
10. http://www.publicintegrity.org/investigations/climate_change/articles/entry/1180/ *The Climate Lobbyists,* by Matthew Lewis. Public Integrity, February 25, 2009.
11. http://www.publicintegrity.org/investigations/climate_change/articles/entry/1280/ *The 'Clean Coal' Lobbying Blitz* by Marianne Lavelle. Public Integrity, April 21, 2009.
12. http://www.hort.purdue.edu/newcrop/duke_energy/Cannabis_sativa.html Handbook of Energy Crops, by James A. Duke. 1983 (unpublished). Posted at Purdue University: Center for New Crops and Plant Products. Accessed October 2011. NB: see also *The Politics of Hemp* by Stuart Jeanne Bramhall, February 9, 2012. The Energy Bulletin: http://www.energybulletin.net/stories/2012-02-09/politics-hemp Accessed June 2012.
13. http://www.cannabis.za.net/biomass.html *Free Biomass Clean Cannabis Energy.* Quoted with permission from *Cannabis: The Facts, Human Rights and the Law* by Kenn and Joanna d'Oudney. The Cannabis Embassy of South Africa. 2011.
14. http://www.hempcouture.ca/inconvenient.html *An Inconvenient Truth About Industrial Cotton.* Hemp Couture, 2007. Accessed October 2011.
15. http://www.cannabis.za.net/biomass.html *Free Biomass Clean Cannabis Energy.* Quoted with permission from Cannabis: The Facts, Human Rights and the Law by Kenn and Joanna d'Oudney. The Cannabis Embassy of South Africa. 2011.
16. http://magblog.audubon.org/ghost-nets-and-mermaid-tears-human-impact-%E2%80%98-last-great-wilderness%E2%80%99 *Ghost Nets and Mermaid Tears:*

Human Impact on "The Last Great Wilderness" by Rachel Nuwer. Audubon Magazine, August 4, 2011.

17. http://www.cannabis.za.net/biomass.html *Free Biomass Clean Cannabis Energy.* Quoted with permission from Cannabis: The Facts, Human Rights and the Law by Kenn and Joanna d'Oudney. The Cannabis Embassy of South Africa. 2011.
18. http://world.std.com/~mica/cftsci.html *Cold Fusion Times: Physics and Material Science Page.* Cold Fusion Times: The journal of the scientific aspects of loading isotopic fuels into materials. Published quarterly. C2011.
19. http://www.tcm.phy.cam.ac.uk/~bdj10/papers/storms/review8.html *Cold Fusion: An Objective Assessment,* Abstract by Edmund Storms, 12/16/01.
20. http://www.newenergytimes.com/v2/reports/LENR-FAQ.shtml#real *Is Cold Fusion Real?* New Energy Times, April 15, 2009.
21. http://www.utexas.edu/lbj/archive/pubs/pdf/20091016_Submitted_LBJ_School.pdf *"Shaping the Energy Technology Transition."* Policy Research Project on Energy Technology Transition: Lyndon Baines Johnson School of Public Affairs. 2009.
22. http://www.newenergytimes.com/v2/reports/LENR-FAQ.shtml#real *Is Cold Fusion Real?* New Energy Times, April 15, 2009.
23. http://www.infinite-energy.com/ Infinite Energy: The Magazine of New Energy Science and Technology, published bi-monthly.
24. http://newenergymovement.org/newenergyp1.php *Open Letter: Universal Appeal for Support for New Energy Science and Technology* by Dr. Eugene F. Mallove. Published May 13, 2004. Posted at New Energy Movement website. Accessed October 2011.
25. http://inventors.about.com/od/sstartinventors/a/Inventors_S.htm *Famous Inventors A to Z.* About.com, Accessed October 2011.

26. http://www.pks.or.at/faq.html *Have Any of Viktor Schauberger's Ideas Ever Been Put Into Practice?* FAQ page, PKS (Pythagoras Kepler System): Society for Promotion of Natural Technology, website. Accessed October 2011.
27. http://wimvos.co.uk/content/view/5/35/ *Copper Plated Plough: Viktor Schauberger* by Wim Vos. August 19, 2004. Website Accessed October 2011.
28. http://www.twentyfirstcenturyhealth.com/?page_id=802 *Viktor Schauberger*, researched and compiled by Denise Turner. Twenty First Century Health, C2011.
29. *Hidden Nature: The Startling Insights of Viktor Schauberger* by Alick Barthlomew with foreword by David Bellamy. Adventures Unlimited Press, 2003. Pages 13-14.
30. http://www.lightnet.co.uk/frontier/viktor.htm Water Ways of Life: Viktor Schauberger. Researched and compiled by Denise Turner. Accessed August 2013.
31. *Hidden Nature,* Bartholomew, page 149
32. http://www.lightnet.co.uk/frontier/viktor.htm opcit, Water Ways of Life.
33. http://merlib.org/node/135 *Viktor Schauberger's Biography* (Timeline). Posted at Modern Energy Research Library website. Accessed October 2011.
34. http://discaircraft.greyfalcon.us/Viktor%20Schauberger.htm *Viktor Schauberger Repulsin A & B (1940-1945)* by Rob Arndt. From an online book titled *Disc Aircraft of the Third Reich (1922 to1945 and BEYOND)* by Rob Arndt. Website accessed October 2011.
35. http://merlib.org/node/135 *Viktor Schauberger's Biography* (Timeline). Posted at Modern Energy Research Library website. Accessed October 2011.
36. http://www.pks.or.at/faq.html *Did Viktor Schauberger Ever Build Flying Saucers?* FAQ page, PKS (Pythagoras Kepler System): Society for Promotion of Natural Technology, website. Accessed October 2011.
37. *Hidden Nature*, Bartholomew, page 243.

38. http://www.spacedaily.com/news/rocketscience-02t.html *Boeing Internal Report Looks At Propellentless Propulsion.* Space Daily, August 1, 2002.
39. http://www.fas.org/man/dod-101/army/docs/astmp/c4/P4K.htm *Chapter IV. Technology Development Army Science and Technology Master Plan. K. Electronic Warfare/Directed Energy Weapons.* Posted by The Federation of American Scientists. Accessed August 2013.
40. http://abcnews.go.com/Politics/navys-laser-weapon-weapon-tomorrow-now-reality/story?id=18908406 *Navy's New Laser Weapon Blasts Bad Guys from Air, Sea.* By Luis Martinez, April 8, 2013. ABC News. Accessed August 2013.
41. http://www.newenergymovement.org/newenergyp1.php *Open Letter: Universal Appeal for Support for New Energy Science and Technology* by Dr. Eugene F. Mallove. Published May 13, 2004. Posted at New Energy Movement website. Accessed October 2011.
42. http://keelynet.com/energy/garrett.htm *The Next Step in Energy Production, Engines and Lighting* by Jerry W. Decker for KeelyNet. Accessed August 2013.
43. http://en.wikipedia.org/wiki/Francois_Issac_de_Rivaz Francois Issac de Rivaz. Wikipedia, Accessed August 2013.
44. http://fuel-efficient_vehicles.org/energy-news/?page_id=44#oiltakeover *How the Oil Companies First Took Over A.K.A. The Oil Age That Never Should Have Been.* Accessed August 2013.
45. http://www.wipo.int/wipo_magazine/en/2007/01/article_0001.html *Green Technologies: Electric Cars with Hydrogen Fuel Cells.* WIPO Magazine. Accessed August 2013.
46. http://money.cnn.com/2013/08/15/technology/fuel-cell-hydrogen.pr.fortune/index.html *Has the Fuel Cell Car's Time Finally Come?* By Brian Dumaine. CNNMoney. Accessed August 2013.
47. http://www.blacklightpower.com/business/facilities/ Business Facilities. Blacklight Power. Accessed August 2013.

48. http://www.blacklightpower.com/business/business-summary/ Business Summary. Blacklight Power. Accessed August 2013.
49. http://www.telegraph.co.uk/earth/energy/renewableenergy/3535012/Ocean-currents-can-power-the-world-say-scientists.html *Ocean Currents Can Power the world, Say Scientists.* Posted on The Telegraph UK, September 6, 2013. Based on a report written by Jasper Copping, November 29, 2009. Accessed August 2013.
50. http://ns.umich.edu/new/releases/6842 *'Fish Technology' Draws Renewable Energy From Slow Water Currents.* University of Michigan News Service. December 8, 2008. Accessed September 2013.
51. http://www.vortexhydroenergy.com/ Vortex Hydro Energy: Clean, Efficient, Powerful – A Breakthrough in Marine Renewable Energy. Accessed September 2013.

Chapter 6

1. http://www.bloomberg.com/apps/news?pid=newsarchive&sid=aXRBOxU5KT5M *Carbon Capitalists Warming to Climate Market Using Derivatives,* by Lisa Kassenar. Bloomberg, December 4, 2009.
2. http://umcp.academia.edu/klaushubacek/Papers/416765/Changing Concepts of LandIn Economic Theory From Single to Multi-Disciplinary Approaches *Changing Concepts of Land in Economic Theory; From Single to Multi-Disciplinary Approaches*, by Klaus Hubacek and Jeroen C.J.M. Van den Bergh. Ecological Economics 56 (2006) 5-27.
3. http://umcp.academia.edu/klaushubacek/Papers/416765/Changing Concepts of LandIn Economic Theory From Single to Multi-Disciplinary Approaches Ibid.
4. http://blogs.ft.com/martin-wolf-exchange/2010/07/12/why-were-resources-expunged-from-neo-classical-economics/#axzz1aQ4nS5m1 *Why were resources expunged*

from neo-classical economics? By Martin Wolf. Martin Wolf's Exchange, July 12, 2010.

5. http://wealthandwant.com/docs/Gaffney_HG100YL_Reconciler.html *Henry George 100 Years Later, The Great Reconciler*, by M. Mason Gaffney, PhD. Posted at Wealth and Want website, accessed October 2011.
6. http://wealthandwant.com/docs/Gaffney_LaaDFoP.html *Land As a Distinctive Factor of Production*, by Mason Gaffney. NB:This working paper later became part of a book titled Land and Taxation, published in 1994 and edited by Nicolas Tideman: list of published works: http://www.masongaffney.org/publications.html
7. Ibid.
8. http://www.masongaffney.org/publications/B5Soil_Depletion_&_Land_Rent.CV.pdf *Soil Depletion and Land Rent* by M. Mason Gaffney chairman of the Dept. of Economics, University of Wisconsin Milwaukee, originally published in The Natural Resources Journal, vol. 4, January 1965
9. Ibid.
10. http://www.nofa.org/tnf/myth.php *The Myth of Our Organic Past* by Jochen Welsch. The Natural Farmer Newsletter, published by the Northeast Organic Farming Association. Article accessed October 2011.
11. http://www.nofa.org/tnf/humus.php *Humus: Still a Mystery* by Paul Sachs. The Natural Farmer Newsletter, published by the Northeast Organic Farming Association. Article accessed October 2011.
12. http://www.nofa.org/tnf/humus.php *Humus: Still a Mystery* by Paul Sachs. The Natural Farmer Newsletter, published by the Northeast Organic Farming Association. Article accessed October 2011.
13. *Unforgiven: the Story of How America Has Exchanged Parity Agriculture for Parity War*. By Charles Walters Jr. Published by Economics Library, Kansas City Missouri in cooperation with Citizens Congress for Private Enterprise. C1971. Page 59.

14. As explained by Alick Bartholomew in *Hidden Nature*, Also see Chapter 3: Climate Change, Carbon Sinks and Sacred Cows: A Better Way to Feed the World? Pages 32 to 41.
15. *Unforgiven*, Walters, page 388.
16. http://www.monetary.org/wp-content/uploads/2013/04/a_program_for_monetary_reform.pdf *A Program For Monetary Reform,* by Paul E. Douglas, Irving Fisher etal. Accessed July 2013.
17. http://www.investopedia.com/terms/m/multipliereffect.asp *Multiplier Effect: Investopedia Explains 'Multiplier Effect.'* Accessed July 2013.
18. http://inflationdata.com/articles/2011/09/17/money-multiplier/ *Money Multiplier* by Tim McMahon, September 17, 2011 quoting *How Wealth Can Simply Evaporate* by Bob Stokes. Accessed July 2013.
19. *100% Money and the Public Debt* by Irving Fisher, pages 10-11. Originally published without copyright notice in the *Economic Forum* Spring Number, April-June 1936 (pages 406-420)
20. http://www.thetwofacesofmoney.com/index.php/Site/NewIMFStudyMakesBigWaves *The Chicago Plan Revisited: an IMF Study* by Jaromir Benes and Michael Kumhof, 2012.
21. http://www.energybulletin.net/node/52591 *Money and the steady state economy*, by Herman Daly. Originally published in the Daly News, April 25, 2010.
22. http://wealthandwant.com/themes/About_HG.html Assorted articles by assorted authors, scroll down to see Mason Gaffney.
23. *Unforgiven*, Walters, pages 54-55
24. http://www.globalresearch.ca/debt-cancellation-in-mesopotamia-and-egypt-from-3000-to-1000-bc/5303136 *Debt Cancellation in Mesopotamia and Egypt from 3000 to 1000 BC* by Eric Toussaint. Center for Research on Globalization, September 4, 2012. Global Research, accessed July 2013.

25. http://www.scribd.com/doc/30493013/American-Empire-Roman-Empire The Role of Christianity: From the Roman Empire to the American Revolution. By John Hudson. Accessed July 2013.
26. *Jefferson and His Time, Volume One: Jefferson the Virginian* by Dumas Malone. Little, Brown and Company, Boston. 1948. page 256.

Chapter 7

1. http://teachingamericanhistory.org/library/document/letter-to-james-madison-17/ Thomas Jefferson to James Madison, September 6, 1789. Accessed July 2013.
2. http://www.jstor.org/pss/4249329 *Bank Notes Will Be But As Oak Leaves: Thomas Jefferson on Paper Money* by Donald F. Swanson. The Virginia Magazine on History and Biography, Vol 101, No. 1(January 1993)
3. http://www.conlaw.org/Intergenerational-II-2-3.htm *Intergenerational Justice in the United States Constitution, The Stewardship Doctrine: II. The Intergenerational Philosophy of the Founders and Their Contemporaries B. Generational Sovereignty and the Land – The Earth as Tenancy in Common – Thomas Jefferson's Usufruct.* Constitutional Law Foundation. Accessed October 2011.
4. http://www.conlaw.org/Intergenerational-II-2-3-1.htm Intergenerational Justice in the United States Constitution, The Stewardship Doctrine: II. The Intergenerational Philosophy of the Founders and Their Contemporaries B. Generational Sovereignty and the Land – The Earth as Tenancy-in-Common – Thomas Jefferson's Usufruct, continued. Constitutional Law Foundation. Accessed October 2011. For more historical context see also: http://ijdb.auzigog.com/author/jefferson-thomas Jefferson,

Thomas submitted by John Davidson of Intergenerational Justice, University of Oregon, September 21, 2010.

5. http://www.yamaguchy.com/library/kellogg/kellogg_index.html *A New Monetary System,* being a collection of the writings of Edward Kellogg, published posthumously by his daughter in 1875.
6. http://michael-hudson.com/2004/01/the-mathematical-economics-of-compound-rates-of-interest-a-four-thousand-year-overview-part-i/ *The Mathematical Economics of Compound Rates of Interest: A Four-Thousand Year Overview Part 1*, by Michael Hudson. January 24, 2004, accessed July 2013.
7. For more insight see *Tobacco Culture: The Mentality of the Great Tidewater Planters on the Eve of the Revolution* by T. H. Breen as well as Jefferson's battle against feudalist laws of inheritance, including primogeniture, escheat and entail.
8. *Unforgiven*, Walters, page 275
9. http://www.dirtyoilsands.org/tarsands *What Are the Tar Sands?* Dirty Oil Sands, accessed June 2013.
10. http://www.canadafreepress.com/2005/deweese060605.htm *UN Targets Your Mayor,* by Tom Deweese, June 6, 2005. Published in Canada Free Press, online. Accessed June 2013.
11. http://www.foodfirst.org/en/Rio%2020%20issues *Green Grabbing our Future: Rio+20 United Nations Conference on Sustainable Development.* By Eric Holt-Giménez, Agnes Walton and Mariagiulia Mariani. Released by the Institute for Food and Development Policy, Food First June 19, 2012, posted at the Food First website. Accessed June 2013.
12. http://iatp.org/documents/who%E2%80%99s-at-the-table-demanding-answers-on-agriculture-in-the-trans-pacific-partnership *Who's At the Table? Demanding Answers on Agriculture in the TransPacific Partnership* by Karen Hansen-Kuhn. Institute for Agriculture and Trade, March 4, 2013. Accessed June 2013.
13. Walters, pages 52-53.

14. http://www.normeconomics.com/index.html Welcome page for NORM: National Organization for Raw Materials. Accessed October 2011.
15. http://www.normeconomics.com/wilken.html Carl H. Wilken, posted at the National Organization for Raw Materials website. Accessed October 2011.
16. http://www.acresusa.com/toolbox/reprints/History%20of%20Money_CW.pdf *The Nature of Money* by Charles Walters. Acres USA Newsletter, November 2003-February 2004. Vol. 33/34.
17. *Unforgiven,* Walters, pages 88-89.
18. *Alexander Hamilton and the Leviathan State* by Vernon Louis Parrington. Excerpted from Main Currents in Economic Thought by Vernon L. Parrington. Harcourt, Brace and Company, 1930. Reprinted with permission in Hamilton and the National Debt, edited with an Introduction by George Rogers Taylor. D.C. Heath and Company, 1950.
19. *The Origins of the Federal Reserve System: Money, Class and Corporate Capitalism, 1890-1913* by James Livingston. Cornell University Press, New York. 1986.
20. Ibid.
21. http://www.thetwofacesofmoney.com/index.php/Site/SlidePresentation See slides 33 and 34 for a description of the two tier system of Fed debt/money creation, and slides 35 through 49 for a discussion of exponential increase affect the economy and the devaluation of the dollar versus what a "public" monetary system would do.
22. opcit, Livingston.
23. http://michael-hudson.com/2013/07/the-bubble-economy-as-a-2-part-play-for-privatisation/ *The Bubble Economy as Part of a 2 Part Play for Privatization* by Michael Hudson. July 4, 2013. Accessed July 2013.

Chapter 8

1. *Banks and Politics in America: From the Revolution to the Civil War*, by Bray Hammond. Princeton University Press, Princeton, New Jersey. C1957. Pages 26-27
2. Ibid, page 15
3. http://rinf.com/alt-news/contributions/what-banks-academics-the-media-and-politicians-dont-tell-you-about-money/4888/ *What Banks, Academics, The Media and Politicians Don't Tell You About Money.* RINF News, November 19, 2008. Page accessed October 2011.
4. http://www.govtrack.us/congress/bills/112/hr2990/text Bill Text, 112th Congress (2011-2012). H.R.2990.IH National Emergency Employment Defense Act of 2011 (Introduced in House – IH)
5. http://www.monetary.org/wp-content/uploads/2010/11/amacolorpamphlet.pdf The American Monetary Act, draft of 2009. American Monetary Institute. Accessed July 2013.
6. http://monetary.org/wp-content/uploads/2011/11/DesignOpenMacro.pdf *Workings of a Public Money System of Open Macroeconomies – Modeling the American Monetary Act* (A Revised Version) by Dr. Kaoru Yamaguchi. Accessed Ju;y 2013.
7. http://www.imf.org/external/pubs/ft/wp/2012/wp12202.pdf IMF Working Paper: The Chicago Plan Revisited, by Jaromir Benes and Michael Kumhoff. August 2012. Accessed July 2013.
8. http://sovereignmoney.eu/100-per-cent-banking-and-plain-sovereign-money *100 Per Cent Banking and Plain Sovereign Money* by Dr. Joseph Huber. Accessed August 2013.
9. http://sovereignmoney.eu/what-would-a-sovereign-money-system-look-like *What Would a Sovereign Money System Look Like?* By Dr. Joseph Huber. Accessed August 2013.

10. http://www.monetary.org/the-1930s-chicago-plan-vs-the-american-monetary-act/2009/08 *The 1930s Chicago Plan and the American Monetary Act* by Stephen Zarlenga, AMI Monetary Reform Conference, Chicago October 2005. Accessed July 2013.
11. http://www.monetary.org/the-need-for-monetary-reform/2009/09 *The Need for Monetary Reform,* American Monetary Institute. September 30, 2009. Accessed July 2013.

APPENDIX I: WHAT IS MONEY?

The world famous Mixt Money Case of 1604, paraphrased: Money is what the government says it is.

Stephen Zarlenga (author of the 2004 book *Lost Science of Money*), per Aristotle, the Mixt Money Case etal: Money is an abstract social power based in law; and whatever government accepts in payment of taxes will be money.

Some of our founders:

Ben Franklin: To remedy such Inconveniences [of barter], and facilitate Exchange, Men have invented MONEY, properly called a Medium of Exchange, because through or by its Means Labour is exchanged for Labour, or one Commodity for another.

James Madison, in a footnote to his 1779 "Report" on money which was later published in Philip Freneau's *National Gazette of Philadelphia* in 1791: "Money is the instrument by which men's wants are supplied, and many who possess it will part with it for that purpose, who would not gratify themselves at the expense of their visible property. Many also may acquire it, who have no visible property."

In the body of his "Report," Madison points out that the key difference between a loan-office certificate and a common bill of credit is the interest charged on the loan-office certificate. Then his zinger: "Instead of paying off the capital to the public creditors, we give them an enormous interest to change the name of the bit of paper which expresses the sum due them; and think it a piece of dexterity in finance, by emitting loan office certificates, to elude the necessity of emitting bills of credit." (This is what Madison, Jefferson and countless others spent the next 40 years fighting. It is what remains with us today, which is to say debt, serving as money.)

Thomas Jefferson, prior to and during the Constitutional Convention, was serving as Ambassador to France (1784-1789). In 1786, he helped French author Jean Nicholas D'meunier write a section on the United States for D'meunier's portion of the *Encyclop die m'thodique.* Replying to questions concerning bills of credit, Jefferson said that wherever state legislatures "laid taxes to bring in money enough for that purpose, and paid the bills punctually . . . paper money was in as high estimation as gold and silver." And then, again paralleling Madison's "Report": "Those who talk of the bankruptcy of the U. S. are of two descriptions. 1. Strangers who do not understand the nature & history of our paper money. 2. Holders of that paper-money who do not wish that the world should understand it."

After his presidency, Jefferson wrote numerous letters to his former Treasury Secretary Albert Gallatin, and his son-in-law John Eppes, then in Congress on the subject of Constitutional money, which he held to be Treasury Notes, bottomed on a tax which would redeem them. These notes he said should provide the circulating medium of the country.

John Taylor of Caroline, writing in 1814, "Money is not intended to create wealth, or the objects of commerce; nor is it able to do either. Its office is to represent and exchange them."

Yours truly: One thing most of us do is to confuse money with wealth. Money is not wealth, it is a receipt for wealth. Wealth itself is best described as anything having exchange value, which sustains life or adds to life's comfort or happiness. John Stuart Mill described wealth as "All things useful and agreeable having exchange value."

Money in other words is the receipt which facilitates exchanges of REAL wealth, that which you and I produce at the local level.

As James Madison, John Taylor, Thomas Jefferson and many others have argued, money is NOT bank paper (or notes) which is really a money substitute created by the banking system; real money is an abstraction, and a creature of law.

So, to put it yet another way, money is simply a device, or ticket system if you will, for the distribution of goods and services, a concept that makes it self-apparent that money must therefore be both an abstraction and a creature of law. Thus, as Dr. Joseph Huber puts it: *Debt-free sovereign money may not be a promise to repay, but it is a promise to be productive, and a promise to keep control of the quantity of money so that there is neither too much nor too little money around in correspondence with actual levels of productivity.*

And Some Quotes on why Gold (or by extension Wheat, diamonds or ANY commodity) Is NOT Money

John Locke in his 1718 *Essay on Money and Bullion*: "Observe well these rules: It is a very common mistake to say that money is a commodity. . . .Bullion is valued by its weight. . . money is valued by its stamp."

Ben Franklin in his *Modest Inquiry into the Nature and Necessity of a Paper Currency*: "Silver and gold . . .[are] of no certain permanent value. . .We must distinguish between money as it is bullion, which is merchandise, and as by being coined it is made a currency; for its value as merchandise, and its value as a currency are two different things."

Hamilton, echoing the argument set forth by Franklin: "It is immaterial what serves the purpose of money, whether paper or gold and silver; that the effect of both upon industry is the same; and that the intrinsic wealth of a nation is to be measured, not by the abundance of the precious metals contained in it, but by the quantity of the productions of its labor and industry."

John Taylor of Caroline, perhaps harking back to the 16th century Potosi disaster "... Specie can transfer wealth from one country to another. If the United States could at pleasure create specie, they might, by a prudent use of such a monopoly, enrich themselves considerably at the expense of the world. . . Overflowing mines of these precious metals, would destroy their utility as a medium of exchange; and confined to one nation, would diminish rather than increase its happiness."

Niall Ferguson, in his popular contemporary book *The Ascent of Money*, after describing in horrifying detail the working conditions of the 16th century Potosi silver mines, restates John Taylor's observation this way: "Like King Midas, the Spanish monarchs of the 16th century found that an abundance of precious metal could be as much a curse as a blessing. The reason? They dug up so much silver to pay for their wars of conquest that the metal itself dramatically declined in value. . . What the Spaniards failed to understand was that the value of precious metal is not absolute. . ."

Ferguson then writes: "The intangible character of most money today is perhaps the best evidence of its true nature. What the conquistadors failed to understand is that money is a matter of belief, even faith: belief in the person issuing the money he uses or the institution that honors his cheques or transfers. Money is not metal. It is trust inscribed. And it does not seem to matter much where it is inscribed: on silver, clay, on paper, on a liquid crystal display. Anything can serve as money, from the cowrie shells of the Maldives to the huge stone discs used on the Pacific islands of Yap. . ."

This concept of money was suggested by James Madison in his "Report" in which he asserts that the value of a nation's money is dependent upon “the credit of the state issuing it”. Madison's discussion of the much maligned Continentals asserts that part of the reason for their decline in value had to do with growing distrust of the public credit, caused in part by “a combination of enemies employing every artifice to disparage [the currency].”

Stephen Zarlenga cites two ancient cases of monetary science from Greece and Rome that supports this idea that "money is not metal, it is trust inscribed":

> "Plutarch describes Lycurgus 8th century BC monetary reform when Sparta's wealth became overly concentrated. He banned using gold and silver and used iron slugs for money. Furthermore those iron pieces were dipped in vinegar while hot, to render them brittle and purposely destroy any commodity value that they had as iron! They received their value through legal sanction. This nomisma system lasted over 3 centuries and Sparta became a premier power. Polybius tells us it faltered when Sparta's involvement in empire regressed her back to gold and silver money and they lost the science of money.
>
> 'Republican Rome based her money on copper, isolating herself from the East and “disenfranchising” the gold/silver hoards and therefore much of the power of the East. Gold could be traded as merchandise; but without the monetary power, the ability of the East to control Rome's money was reduced and she had a better chance to control her destiny. Roman Nomisma, were bronze discs legally valued far above their commodity content through the law.”

On the question of whether gold is money, Alexander DelMar relates the following in his treatise *The Science of Money*:

> ". . .The Mixt Moneys case [of 1604] decided that Money was a Public Measure, a measure of value, and that, like other measures, it was necessary in the public welfare that its dimensions of volume should be limited, defined and regulated by the State. The whole body of learning left us by the ancient and renascent world was invoked in this celebrated dictum: Aristotle, Paulus, Bodin and Budelius were summoned to its support; the Roman law, the common law and the statutes all upheld it; "the State alone had the right to issue money and to decide of what substances its symbols should be made, whether of gold, silver, brass, or paper. Whatever the State declared to be money, was money" . . . and observes that By monopolizing this [gold] commodity the moneyed classes have got Nature by the throat and the community under their heels. . . Compared with this process, usury is mere child's play."

The more contemporary problem that allows the myth that money must be based on some commodity such as gold or other type of collateral to persist is discussed in brief by Alexander Del Mar: "As a rule political economists . . don't take the trouble to study the history of money; it is much easier to imagine it and to deduce the principles of this imaginary knowledge."

1935 statement by Frank Vanderlip, who served as Assistant Secretary on the Treasury under William McKinley and later as President of the National City Bank of New York, and who was one of the seven original authors of Federal Reserve Act:

"What is it we want of our currency? . . . We want a dollar that will, in the language of the President, "not change its purchasing and debt paying power during the succeeding generation" . . . What are the points to be taken into consideration? First and foremost, that Congress should assume the responsibility laid on it by the Constitution of regulating the value of money. We now know that a given weight of gold is not an unchanging standard of value. That fact is dawning slowly on the most conservative and obstinate minds."

INDEX

Acid Rain Program 31, 47
actual value (of money) **161-162, 166-167, 173, 184, 187, 196, 218**
see also Alexander Hamilton's Financial plan, Bank of the United States (First), centralized control (of monetary policy), dollar devaluation, exponential laws of increase, Federal Reserve System, fractional reserve deposit expansion, legal value (of money), money multiplier, usufruct
Agenda 21 .... **174-175, 178**
see also free trade, United Nations
Agreement on Agriculture **171**
see also free trade
agricultural inputs **57, 59, 75**
agriculture **2, 10, 15-16, 20, 25-26, 29, 33, 37-39, 41, 46, 57, 60-61, 63-65, 68, 70, 77-78, 93, 116, 130, 137-140, 143-144, 171, 213**
see also CAFO operations, carbon depleted soil, carbon sequestering agriculture, carbon sinks, Codex Alimentarius, factory farms, free trade, green revolution, humus, industrial agriculture, monocrop agriculture, oil-seed tree plantations, peak soil, perennial agriculture, plantation-style agriculture, regenerative agriculture, soil carbon, soil fertility, soy highway, tree plantations, usufruct, waste lagoons, water use
agrofuels **81, 91-96, 100-101, 103-104, 106, 177**
see also algae, biofuel, cellulose, hemp
air particulates **91**
air pollutants **30, 47-48, 51**
see also negative externalities, pollutants, pollution
Albrecht, William **45**
ALEC (American Legislative Exchange Council) **173**
Alexander Hamilton's financial plan **145-148, 185-188**
see also actual value (of money), Bank of the United States (First), centralized control (of monetary policy), dollar devaluation, exponential laws of increase, Federal Reserve System, fractional reserve deposit expansion,

legal value (of money), money multiplier, usufruct
algae.. 6, 56, 96-97
see also agrofuel, biofuel, cellulose, hemp
Bank of the United States (First) 145-146
see also actual value (of money), Alexander Hamilton's financial plan, centralized control (of monetary policy), dollar devaluation, exponential laws of increase, Federal Reserve System, fractional reserve deposit expansion, legal value (of money), money multiplier, usufruct
Benes, Jaromir.. 203
Berry, Wendell 138
biodegradable plastics.. 105-106
see also hemp
biofuels 8, 21, 36-37, 41, 59-60, 71, 80-82, 89-97, 100, 103-104, 177
see also agrofuels, algae, cellulose, hemp
boreal forests 67, 172
see also NAFTA, peatlands, rainforests, wetlands
CAFO (concentrated animal feeding operations)...................25-26, 59
see also factory farms
cap and trade... 27-33, 36, 52, 80, 129
see also carbon business, derivatives, carbon markets carbon pricing, carbon tax
carbon.................................. 1, 5-6, 11, 16-17, 19-24, 26-34, 36-42, 47, 52, 55- 56, 58, 68, 73-74, 77, 96, 129, 139
see also CO2, fertilization effect
carbon cycle 16-17, 38
carbon depleted soil 6, 24, 41, 65, 73, 218
see also fertilization effect
carbon derivatives .... see derivatives
carbon markets...............9, 28-29, 34, 36, 73, 85, 93, 100-101, 130, 177
see also cap and trade, carbon business, derivatives, carbon pricing, carbon tax
carbon monoxide 27, 30, 47, 91
carbon pricing... 27, 30, 52, 78
see also cap and trade, derivatives, carbon markets, carbon tax
carbon sequestration.. 77, 82, 84, 139
carbon sequestering agriculture 4, 26, 28-29, 103

see also intensive grazing management, perennial agriculture, regenerative agriculture
carbon sink....................................21, 34, 36-38, 60, 69, 71, 73, 129
see also agrofuels, oil-seed tree plantations, plantation style agriculture
carbon tax 28, 30-31, 33-34, 94
see also cap and trade, derivatives, carbon markets, carbon pricing
cellulose 71, 96-97
see also agrofuels, algae
centralized control (of monetary policy) 168, 186, 189
see also Alexander Hamilton's financial plan, exponential laws of increase, Federal Reserve System, fractional reserve deposit expansion, money multiplier, usufruct, vertical and horizontal integration
chlorofluorcarbons 54
see also greenhouse gases, nitrous oxide
Chossudovsky, Michel....................................169-170
Clean Air Act 30-31, 47-48, 84,
clean energy 11, 79-82, 85, 91, 95, 100-103, 127-128
see also agrofuels, coal, hemp, natural gas, new energy, nuclear power, renewables, solar, wind
clear-cutting....................................38, 42-43, 61, 68, 81, 92
see also land use
climate change.........1, 3, 7-8, 16-18, 20, 24, 26-27, 29, 34, 44-44, 46, 56, 60, 78, 90-91, 94, 129, 143, 172, 218
see also global warming
climate remediation....................................4-5, 10, 34-35, 38, 43, 47, 52, 94, 120
see also geoengineering, weather modification
CO2 ....1, 4, 6, 8-9, 11, 17, 19-20, 24-26, 28, 30-31, 33-34, 36, 39-40, 43-44, 48, 53-55, 60, 68-69, 73, 78, 80-82, 85, 89,
92, 96-97, 101, 129
see also carbon, fertilization effect
coal 25, 38, 53, 79, 81-82, 89, 90, 94, 103, 123
Codex Alimentarius....................................178
see also free trade
cold fusion 107-110, 112, 120

see also Mallove, Eugene, new energy, Schauberger, Viktor
common air pollutants....................30, 33, 47-48, 54
see also air pollutants, negative externalities, pollution
concentrated animal feeding operations....................see CAFO
crop yields....................6, 40-43, 53, 64
see also humus, soil carbon reserves, soil fertility
Daly, Herman....................151-152
Daniel, T. Cushing....................168, 198
degraded soils....................6-7, 24, 61, 78
see also carbon depleted soils, soil carbon
DelMar, Alexander,....................191
derivatives....................32, 34, 57-58, 101-102, 130, 176, 178, 187
desertification....................57-58
see also Dust Bowl, land use, Mesopotamia
dioxin....................46
dollar devaluation....................165
see also actual value (of money), exponential laws of increase, legal value (of money)
Dust Bowl....................22-25
see also land use, Mesopotamia
earth's climate system....................5, 19, 35, 43
economics....................67, 124, 136, 139, 142, 154, 188-189, 234 263-264, 267
economists....................8, 11, 27, 52, 132, 134-135, 144-145, 154, 163-164, 167, 182, 185, 202-203
see also Benes, Jaromir; Chossudovsky, Michel; Daly, Herman; Fisher, Irving; Gaffney, M. Mason; George, Henry; Huber, Joseph; Hudson, Michael; Kumhoff, Michael; Walters, Charles, Yamaguchi, Kaoru
electric cars....................122-124
energy....................11, 32-38, 55, 59, 79 -129, 129. 173, 175, 215
see also clean energy, cold fusion, free trade, hemp, new energy, renewables
energy sprawl....................83, 89
see also Nature Conservancy
environmental degradation....................46
see also land use, negative externalities, pollution

exponential laws of increase 163-166, 188, 200, 209
see also Hudson, Michael, money multiplier
factory farms 59
see also CAFO (concentrated animal feeding operations)
Federal Reserve System 146, 185-189, 210-211
see also actual value (of money), Alexander Hamilton's Financial plan, Bank of the United States (First), centralized control (of monetary policy), dollar devaluation, exponential laws of increase, fractional reserve deposit expansion, legal value (of money), money multiplier, usufruct, vertical and horizontal integration
fertilization effect 6, 39, 43
see also CO2, soil carbon
fertilizers 6, 23, 25, 30, 36, 42, 44, 47-48, 54-59, 65-66, 75, 92, 105, 108, 177
First Bank of the United States..........see Bank of the United States (First)
Fisher, Irving 148, 150-151, 182, 202
food quality... 11, 19, 41, 43, 47, 76, 212-213
food security.. 41, 43
fracking 83-84
see also NAFTA
fractional reserve deposit expansion 149-150, 186
see also actual value (of money), legal value (of money), money multiplier, usufruct
free trade 46, 63, 103, 168-171, 173-174, 178-179, 181, 189, 210, 213, 223
fuel hydrocarbons 46
Gaffney, M. Mason 132-133, 145, 153, 160
see also humus, land rent, usufruct
gas 25, 38, 49, 79-85, 89-91, 99, 106, 115, 121-123, 215
GATT 171
see also free trade
genetically modified crops 42, 95
see also green revolution
geoengineering 5, 34-35
see also climate remediation, weather modification
George, Henry 131, 133, 135, 152, 161, 167, 187

geothermal 81-82, 89
GHG .... see greenhouse gases
global warming 1, 3-7, 15-17, 19, 25, 27, 35, 41, 44, 54, 60, 94, 101-102
  see also climate change
globalization 167-169, 175
  see also free trade
glyphosate (Round Up) 39
GMO foods 44
  See also green revolution
Green Economy 177
green revolution 4-43, 45, 57, 61, 63, 65-66
green technologies 82
greenhouse gases 16, 18, 25, 35, 40, 48, 86-87, 91, 101, 215
Greenpeace 90-91, 101, 103
halocarbons 17, 54
Hamilton, Alexander 145-148, 185-186, 195
  see also Alexander Hamilton's Financial Plan, Bank of the United States (First), centralized control (of monetary policy), Federal Reserve System, usufruct
hemp 104-107, 122
  see also biofuels
Huber, Joseph 205
Hudson, Michael 154, 163, 170, 189
humus 11, 20-23, 74-75, 77, 137-140, 144, 218
  see also soil carbon reserves, usufruct
hunger 6, 60, 62-64, 73, 181
  see also food quality, malnutrition, micronutrient depletion, soil fertility, soil quality, world hunger
hydraulic fracturing .... see fracking
hydroelectric dams 70, 81, 115
hydrogen fuel cell cars 123-24
hydrogen powered cars 123-124
hydropower 81-82, 84, 88
indigenous populations 7, 35, 60. 67-68, 92, 103
industrial style agriculture 10, 16, 23, 25-26, 40-41, 44, 46, 56-57, 60, 68, 70, 77, 139

see also CAFOs, green revolution, monocrop agriculture, plantation style agriculture, soy highway, tree plantations
intensive grazing management 74-75
see also regenerative agriculture, carbon sequestering agriculture
irrigation 6, 23-24, 57-58, 65, 105, 154-155
see also Ogalla Aquifer, water use
Jefferson, Thomas 11, 141, 145, 151, 157-158, 166, 183, 192, 190, 201, 218-219
see also usufruct
Keystone XL Pipeline 100, 172-173
Kumhof, Michael 203
lagoons .... see waste lagoons
land rent 135-136, 138, 144
see also Gaffney, M. Mason
land use 1-5, 7-8, 10-11, 15-16, 22-23, 25, 27, 28-29, 33, 36-37, 41, 44, 52, 66, 76, 85, 91, 94, 114, 129-130, 150, 218-220
legal value (of money) 161-162, 167, 196, 218
see also actual value (of money), dollar devaluation, fractional reserve deposit expansion, money multiplier
legalized poverty 64
see also free trade
Mallove, Eugene 110, 120, 123, 128, 134
see also cold fusion, new energy, Schauberger, Viktor
malnutrition 62-64
see also food quality, hunger, micronutrient depletion, soil fertility, soil quality
man-made clouds 35
see also climate remediation, geoengineering, weather modification
man-made earthquakes 84
see also carbon sequestration, fracking, hydroelectric dams
manure lagoons .... see waste lagoons
see also CAFO (concentrated animal feeding operations)
mermaid tears 106
see also biodegradable plastics
Mesopotamia 2-3, 7-8, 16, 25, 64, 67, 72,

102, 114, 154-155, 157, 160, 183
see also Dust Bowl, Hudson, Michael, land use
methane 17, 25-26, 54, 59, 69-70, 92
micronutrient depletion 60
see also see also food quality, malnutrition, soil fertility, soil quality
monetary policy........ see Alexander Hamilton's financial plan, Federal Reserve System, Mesopotamia, Jefferson, Thomas, Daly, Herman, Daniel, T. Cushing, Fisher, Irving, Benes, Jaormir, Kumhoff, Michael, Walters, Charles, Yamaguchi, Kaoru, Zarlenga, Stephen, usufruct
money multiplier 149-150
see also actual value (of money), dollar devaluation, fractional reserve deposit expansion, legal value (of money)
monocrop agriculture........ 25, 33, 38-39, 63, 139
see also carbon sinks, green revolution, industrial style agriculture, oil-seed tree plantations, plantation style agriculture, soy highway, tree plantations
NAFTA........ 171, 173
see also free trade
Nature Conservancy........ 89, 91-92
see also energy sprawl
negative externalities 27, 44, 52, 84-85, 89, 94
see also cap and trade, carbon tax, pollution
new energy........ 109-112, 120, 123-126, 128, 215
see also cold fusion, Mallove, Eugene, Schauberger, Viktor, VIVACE
nitrogen 6, 17-18, 20, 23-24, 30, 32-33, 44, 46-47, 49-56, 59, 65-66, 91, 96 137, 172
nitrogen cycle........ 18
nitrogen loading........ 18
nitrous oxide........ 17, 24-26, 33, 44, 54-55, 92, 96
see also chlorofluocarbons
nuclear power........ 42, 81, 86-89, 90, 120
nutrient requirements (plants)........ 18
see also carbon cycle, carbon depleted soil, fertilization effect, fertilizers, humus
Ogalla Aquifer 23

see also water use
oil 8, 25, 37-38, 59-60, 66, 69, 79, 80, 85, 89, 90-93, 95-96, 99-100, 103-104, 123, 172, 176, 215,
oil-seed tree plantations 8, 37, 91
see also carbon sinks, agrofuels
oxygen.................................6-7, 17-19, 21, 49-50, 55-56, 68-69, 103
ozone 17, 30, 48-56
PCBs...42, 46
peak soil 33, 64, 96, 98, 104, 106
peatlands 38, 67-69, 81
see also boreal forests, rainforests, wetlands
Perennial agriculture... 77-78, 139
see also carbon sequestering agriculture, regenerative agriculture
pesticides................23, 25, 30, 36, 44, 46, 48, 55, 57-58, 92, 105, 108
plantation style agriculture 26, 37-39, 55-56, 60, 63, 70, 77, 213
see also agrofuels, carbon sinks, industrial style agriculture, monocrop agriculture, oil-seed tree plantations, soy highway, tree plantations
plastics.......................................see biodegradable plastics, mermaid tears
pollutants 10, 26-27, 30-31, 33, 42, 48, 51, 54, 86, 91
see also negative externalities
pollution 1, 5, 8-9, 26-27, 30-31, 34-35, 38, 43, 52, 60, 63, 75, 81, 85, 89, 94, 106, 133, 177, 215
see also negative externalities
propaganda 15, 87, 93, 184, 190
rain forests..................7-8, 21, 52, 56, 68-70, 81, 85, 92, 103, 129, 215
REDD+ 177
see also free trade
regenerative agriculture............................. 67, 77-78, 138-139, 213, 215
see also carbon sequestering agriculture, intensive grazing management, perennial agriculture
renewable energy 31, 79-82, 85, 88, 90, 93, 95, 100, 102, 126, 128, 175
see also agrofuels, clean energy, hemp
reserves (soil and monetary) 9-10, 22, 26, 38, 138, 143-144,186, 196, 201, 211, 218
see also Fisher, Irving, Gaffney, M. Mason, Jefferson, Thomas,

Schauberger, Viktor, Walters, Charles, usufruct
Rio 176
see also free trade
Round Up .... see glyphosate
Salatin, Joel 74-77, 138
see also carbon sequestering agriculture, intensive grazing management, perennial agriculture
Schauberger, Viktor 112-118, 127, 143
see also cold fusion, new energy, reserves (soil and monetary)
Simons, Henry 202
single-substance science 43
soil 1-2, 6-7, 10-11, 17, 20-27, 33, 37-38, 40-45, 52, 55-58, 61, 63-65, 67-68, 71-74, 77-78, 85, 92, 94, 96-97, 105-107, 112, 114, 135-139, 144, 159, 161, 172, 212, 218
see also carbon cycle, fertilization effect, humus, peak soil, soil fertility
soil degradation 20, 43-44, 46, 56, 62, 72, 85, 114
see also carbon, fertilization effect, humus, peak soil
soil fertility 20, 23, 135-137, 139-141, 144
see also carbon, fertilization effect, humus, peak soil
soil quality 41, 45, 47, 56, 77, 105-106, 212
solar 4, 35, 79, 81-82, 86, 88-90, 107, 125-126, 175
soy highway 70
see also industrial style agriculture, monocrop agriculture, plantation style agriculture, tree plantations
subsidies 33, 62-63, 94, 146, 213
see also free trade
sulphur dioxide 30-31, 44, 47, 52-53, 91, 172
superbugs 95
superweeds 95
tailing ponds 171-172
see also NAFTA
tar sands mining 81, 103, 171-172
TPP (TransPacificPartnership) 179-181
see also free trade
TTIP (Transatlantic Trade and Investment Partnership) 179
see also free trade

transfer of ownership........163-165
see also exponential laws of increase, Hudson, Michael, legal value of money
tree plantations........8, 21, 34, 37-39, 69-72, 76, 95
see also carbon sinks, industrial style agriculture, monocrop agriculture, plantation style agriculture, soy highway, tree plantations
TRIPS Agreement........171
see also free trade
trout turbine technology........127
see also cold fusion, new energy, Schauberger, Viktor, VIVACE
United Nations........57, 174, 178
see also Agenda 21, free trade
usufruct........2-3, 135, 155-162, 167-168, 170, 181, 184, 187, 195-196, 200-201, 218
see also Gaffney, M. Mason, Jefferson, Thomas
vertical and horizontal integration........185, 188
see also centralized control (of monetary policy)
VIVACE........127-128
see also new energy, Schauberger, Viktor
volatile organic compounds (VOC)........47, 52
Walters, Charles........140-145, 148-153, 158, 161, 168, 201-202
see also Alexander Hamilton's financial system, free trade, usufruct
waste lagoons........25-26, 59
water use........58,72, 77, 81-82, 84, 92, 96-97, 108, 172, 208
water vapor........17-19, 22, 35
see also greenhouse gases
weather modification........4-5, 43, 103, 129
see also climate remediation, geoengineering
wetlands........60, 62, 68
see also boreal forests, Mesopotamia, peatlands, free trade, NAFTA
wind........4, 22, 69, 79, 81-82, 86-90, 103, 107, 125, 175
see also clean energy, renewables
world hunger........60, 62-64, 73, 181
see also food quality, hunger, malnutrition, micronutrient depletion, soil fertility, soil quality

WTO (World Trade Organization) 171, 174, 178
see also free trade
Yamaguchi, Kaoru..202-203
Zarlenga, Stephen 201, 207, 216

ABOUT THE AUTHOR

Geraldine Perry holds a Masters Degree in Education, and is a Certified Natural Health Consultant. Reflecting her lifelong interest in reading and research, she minored in library science for her Bachelor of Arts degree, and served as a district wide school library supervisor for eleven years. Her first book, *The Two Faces of Money*, was co-authored with Ken Fousek. She has written articles for *Dissident Voice*, *Online Journal* and other online journals. She also does small group presentations on health and monetary topics. In her spare time she enjoys reading, taking long walks, playing with her cat Ali and picking raspberries in summer. She actively supports and is personally acquainted with the farmers who supply her and her locally engaged food community with some of the best, nutrient dense food on the planet. She is the mother of two grown sons and resides with her husband and her furry friend Ali Cat in the Southwest Suburbs of Chicago.

Made in United States
Orlando, FL
04 November 2026

90205150R00171